Small-Scale Renewable Energy Systems

Small-Scale Renewable Energy Systems

Independent Electricity for Community, Business and Home

Sven Ruin

TEROC AB, Köping, Sweden

Göran Sidén

Halmstad University, Halmstad, Sweden

CRC Press
Taylor & Francis Group
Boca Raton London New York Leiden

CRC Press is an imprint of the
Taylor & Francis Group, an **informa** business

A BALKEMA BOOK

CRC Press/Balkema is an imprint of the Taylor & Francis Group, an informa business

© 2020 Sven Ruin & Göran Sidén

Typeset by Apex CoVantage, LLC

Library of Congress Cataloging-in-Publication Data

Names: Ruin, Sven, author. | Sidén, Göran, author.
Title: Small-scale renewable energy systems : independent electricity for community, business and home / Sven Ruin, TEROC AB, Köping, Sweden, Göran Sidén, Halmstad University, Halmstad, Sweden.
Description: Leiden, The Netherlands : CRC Press/Balkema is an imprint of the Taylor & Francis Group, an Informa Business, [2019]
Identifiers: LCCN 2019020881 (print) | ISBN 9780367030971 (hbk : alk. paper) | ISBN 9780429020391 (ebook)
Subjects: LCSH: Small power production facilities. | Renewable energy sources.
Classification: LCC TK1006 .R85 2019 (print) | LCC TK1006 (ebook) | DDC 621.042—dc23
LC record available at https://lccn.loc.gov/2019020881
LC ebook record available at https://lccn.loc.gov/2019981541

Published by: CRC Press/Balkema
            Schipholweg 107c, 2316 XC Leiden, The Netherlands
            e-mail: Pub.NL@taylorandfrancis.com
            www.crcpress.com – www.taylorandfrancis.com

ISBN: 978-0-367-03097-1 (Hbk)
ISBN: 978-0-429-02039-1 (eBook)

DOI: https://doi.org/10.1201/9780429020391

In memory of Lars Engvall.
Without him, this book would not exist.

Contents

Chapter 1

Introduction

The generation of electricity from renewable energy sources has undergone rapid development in recent years. Globally 14 GW solar photovoltaics were installed as of 2008. In 2018, installations reached 512 GW, according to preliminary statistics. This is an increase of 37 times in 10 years, an incredible development. Solar electricity generation was about 2.2% of the world's electricity needs (Fig. 1.1).

Total installed wind power in the world was 121 GW in 2008. In 2018, the installed power had increased to 600 GW – a fivefold increase in 10 years. Wind power generated 6% of the global electrical power supply. In total, renewable energy provided more than 27% of the global needs for electricity.

Some other renewable energy sources, such as hydropower and bioenergy, have played a large role for a long time and are often a great complement to more fluctuating sources such as solar and wind. New developments are making it possible to have an increasing share of renewable sources in the energy mix. For example, in remote locations biofuels can be used to enable existing diesel engines to switch over to provide renewable generation. Of special interest for remote locations is also that renewable energy can be used on a very small scale (Fig. 1.2).

Figure 1.1 House with solar photovoltaic system including battery storage.
Source: Courtesy of Whidbey Sun & Wind LLC, Washington, USA.

Figure 1.2 The submerged Ampair UW100 generator can provide 100 W at 12, 24 or 48 V DC in combination with a battery (which can also be charged from other sources). It can be used, for example, for run-of-river hydropower, on towed equipment or to capture tidal currents.

Source: Seamap (UK) Ltd.

Another interesting development is the rapid growth and price drop of batteries, in particular lithium-ion, because of the mass production of electric vehicles (EVs). Electricity storage costs will probably continue to decrease, for example, because of the second-life of EV batteries. Batteries that can store electricity – for example, from renewable sources – at low prices open up interesting possibilities in both developing and developed economies. In areas without power grids, local systems for houses or villages, for example, will be more viable than ever before. However, especially in off-grid areas, many are unaware of these new opportunities. Some challenges can be lack of skilled technicians and standard solutions.

Electrification can be done in new ways, such as using EVs for vehicle-to-home, without expensive power grids. Buying, for example, a second-hand EV can often be a cost-effective way to buy electricity storage, and then you also get a car with it, which can transport more than people. EVs provide an opportunity for a "grid on wheels" that can carry significant amounts of energy. Even where large power grids exist, the possibility of building houses without connection to the grid can provide a new situation in the electricity market. The monopoly situation of grid owners is challenged. Many power utilities are aware that a paradigm shift is taking place, and some are looking into how they can be part of that future. For example, E.ON in Simris in southern Sweden is testing a microgrid.

In transportation, the ongoing small-scale energy revolution is opening up new opportunities to make fuel locally, which can be used for many types of vehicles. For example, biogas is produced by many municipalities to be sold at local dispensers, to fuel city buses or to drive generators. Small fuel cells can be used, for example, with upgraded biogas, to produce electricity more efficiently than has ever been possible with an engine. EVs can be charged from local renewable sources (Fig. 1.3), also in off-grid systems with sufficient capability. Hydrogen enables fuel to be made locally from electricity and water, to be stored in large quantities at a lower cost than a huge battery bank, and to be converted back to electricity.

Renewable energy is important for global climate concerns. Renewable energy sources provide smaller emissions of greenhouse gases. So far, environmental concerns have been a major driving force in the development of renewable energy. A change now is that renewable energy sources also have become more economically competitive. Reports from Ecofys (for the EU) and the US Energy Information Administration (EIA) show that the cost of electricity from land-based wind power is comparable or lower than that from fossil sources and nuclear power. Electricity from solar photovoltaics is not far behind. In areas with good access to solar radiation, it can already provide the cheapest new power generation.

Even more competitive is usually the efficient use of energy. Smart use of energy usually goes hand in hand with renewable energy and is of vital importance for the huge challenge facing our society to achieve sustainable development. On the local level, it can, for example, make an off-grid system much more affordable, because generation does not have to be oversized.

Energy security is another driver. Renewable energy reduces reliance on imported energy. If done in a smart way, distributed renewable energy can also contribute to grid reliability and to the better handling of emergency situations. One example of the latter is a study and

Figure 1.3 Inauguration in 2015 of Giraffe 2.0, which is a wind and solar power station by InnoVentum. It can be used, for example, in combination with charging electric vehicles. The basic configuration is for on-grid, but could also be made for off-grid.

Source: S. Ruin.

field tests done by Spirae and Energynautics for eneginet.dk in Denmark regarding coordinated control of local assets. They found that in the event of a transmission system emergency, local distribution networks (60 kV and below) could be rapidly isolated from the transmission network (150 kV and above) and operated autonomously using local resources. This "safe island" operation would reduce the impact on electricity consumers and contribute to a more rapid recovery from the emergency.

An individual home or business owner can provide for their own energy security in a more economical way than before, for example with a grid interactive renewable energy system that normally sells surplus generation, perhaps also participates in grid frequency regulation and, in case of a grid failure, provides grid backup at least to the most important local loads. Such systems can reduce the burden on the grid during a shortage, thereby reducing the risk of grid collapse. Should the grid fail, being able to generate your own electricity can be very important, for example to drive a water pump.

There are many reasons why people want independent power systems. Tiny systems with solar photovoltaics or small wind turbines, batteries, charge controller and perhaps inverter for AC are used in many holiday homes, caravans and pleasure boats. They often provide the most fundamental needs for lighting, refrigerators and supply to electronics such as computers, mobile phones, radio and TV. With the ongoing development of for example batteries, they can meet even more needs in a fully acceptable manner, also in mobile applications.

For many holiday homes it is actually economical not to connect to a power grid. For users who need only a small amount of electricity, the high fixed costs of the grid can be especially prohibitive. If the distance is long to the existing grid, the case for grid connection becomes even weaker.

In remote villages or areas that are difficult to reach with electricity grids, for example on islands, it is often a necessity to rely on small independent systems instead of national electricity grids. Where they have small isolated electricity grids, they have so far often been supplied with diesel-powered power plants. Such power supply is often both insecure and expensive per kWh. With local plants that produce renewable energy, they can get a better and cheaper supply.

In developing countries, for example in sub-Saharan Africa and in South Asia, many people still lack access to an electricity supply. It is estimated that more than one billion people are in this situation. In many such areas, small local, renewable energy solutions are being developed as a more viable way to increase access to electricity compared with national electricity grids. For example, the World Bank [1] has reported on investments in countries such as Bangladesh, Ethiopia, Kenya, Haiti and Bolivia. A total of US$ 1.8 billion has been invested over the past eight years. Also in large economies like Brazil where most people have access to electricity, there is a disadvantaged minority who don't. One approach to changing this is through training of local actors [2].

Electricity access is not a binary question. An additional one billion people are estimated to live with unreliable or insufficient electricity service. Electrical safety is also an issue in many places.

Our society is becoming increasingly dependent on a functioning electricity supply for household, business and many other needs. For example, water supply, fuel dispensers, internet communication and modern payments will normally not work without electricity. This dependence is actually a danger. Therefore, many people want the opportunity to be independent and have control over the electricity supply. Smart-grid solutions can also provide new opportunities for managing the electricity supply, also locally, but if designed in an unsuitable way, they can contribute to the danger (perhaps they can also be hacked).

An important development is also improved understanding of the renewable energy resources. In many cases, such knowledge is becoming more accessible to anyone. For example, the International Renewable Energy Agency, IRENA, has on the Internet published global resource maps of average solar irradiation and wind climatology [3], which are free to access. Knowledge is also shared directly between interested stakeholders on online platforms such as Energypedia (energypedia.info).

To summarize: While environmental concerns have been driving much of the development of renewable energy, we are now in a situation where the economic aspects are emerging as drivers. With the new, cheaper renewable energy, the improved storage possibilities, smart controls and energy-efficient technology, the costs can be lower in the long run – both for those who already have electrical power and those who don't. However, we also should not underestimate the force inherent in the idea of being completely self-sufficient. Off-the-grid (OTG) has become the symbol of a system and lifestyle designed to meet needs without the support of remote infrastructure, such as a power grid. Off-the-grid homes aim to achieve autonomy; they do not rely on community services for water supply, sewage, gas, electricity or similar services. This brings opportunities to provide energy for all and empower local people.

References

[1] Riccardo Puliti, "Off-grid bringing power to millions", *World Bank blogs*, 2018-02-26. [Online]. Available from: http://blogs.worldbank.org/energy/grid-bringing-power-millions [Accessed 2019-05-27].

[2] Instituto Peabiru, "Light for a Better Life". [Online]. Available from: https://peabiru.org.br/light-for-a-better-life/ [Accessed 2019-05-27].

[3] IRENA – International Renewable Energy Agency, "Global Atlas for Renewable Energy". [Online]. Available from: https://www.irena.org/globalatlas/ [Accessed 2019-05-27].

Chapter 2

Electricity generation

2.1. Solar power

The most common technology for solar electricity generation is photovoltaic (PV) cells, which convert the photons, the energy carriers in sunlight, directly into electric power. The building blocks of PV technology are truly small scale, but can be combined to make large systems. The solar cell that supplies a watch generates power equal to only a few milliwatts, while the largest modules generate some hundred watts. Big solar power plants are composed of thousands of modules. But technically speaking, a solar module on a street light is as effective as a large field with thousands of solar modules, and when the electricity is produced close to consumers, the losses in the transfers are small.

Solar PV can be used for small-scale local application as well as a base for a future electrical system with decentralized power supply with millions of producers and consumers who buy and sell energy.

2.1.1. PV technology

The most common solar cell (Fig. 2.1) is made of an approximately 0.2-millimeter-thick plate of silicon (Si), a semiconductor material. When used for photovoltaic, small amounts of other atoms are added; the silicon is doped. Doping with phosphorus gives n-silicon (negative charge carrier). If we dope with boron we get p-silicon (positive charge carriers).

Figure 2.2 shows the structure of a solar cell. In the front there is a top contact consisting of a metal network that collects the charges formed by the solar radiation on the solar cell. The network collects all charges on the top, but at the same time covers as little area as possible, so as not to prevent photons from reaching the silicon crystal.

Under the network, there is a relatively thin layer of n-doped silicon. Below this is a slightly thicker layer of p-doped silicon. In the bottom of the cell, there is a full metal layer, a bottom contact that collects and conducts the charges from the bottom. Thus, the load receiving the energy is connected in between the front and rear connector.

The PV cells are commonly serially connected, mounted in an aluminium frame and covered by tempered glass to form a module. A slightly flexible module can be obtained by covering it with plastic; these applications are popular in boat contexts. Modules can be connected in series or parallel to form an array. However, it is not recommended to connect too many cells or modules in parallel, in order to handle the possible fault of defect or short-circuit.

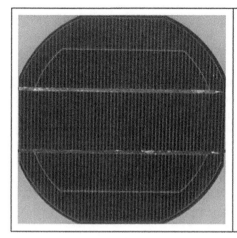

Properties of a typical solar cell

Size: 1 dm²
Voltage: 0.5 V (DC)
Current: ~2 A
Power: ~1 W
Efficiency: 15-20%
Produce: 1-3 kWh / year
Price (in a module): $0.2 - 0.4 €/W_p$

(These are the typical values of today's solar cells. Each type has data that are more accurate.)

Figure 2.1 Solar cell – the truly small-scale electricity producer.

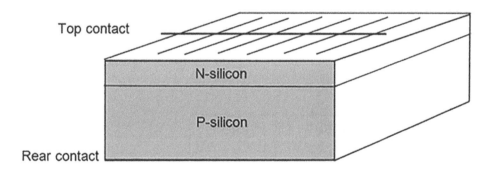

Figure 2.2 The structure of a solar cell.

2.1.2. How it works

The relatively firmly bound electrons in silicon atoms' outermost electron shell is said to be in the "valence band." If they receive a certain quantity of energy, that can be absorbed, then the electrons are raised to a higher energy level called the conduction band. The electrons become excited and can move in the silicon crystal. The energy that is necessary to move an electron from the valence band to the conduction band is called the band gap, for silicon the band gap is 1.1 eV. Approximately 23% of sunlight photons are above the required energy level of 1.1 eV.

When a photon hits the valence electrons in the solar cell, three situations can arise. Figure 2.3 illustrates the three cases:

• Energy content that is less than the band gap cannot be used at all; the electron remains in the valence band
• Energy content equal to the band gap, 1.1 eV, excites the electron, and it becomes mobile in the conduction band

- Greater energy content than the band gap also provides a moving electron, but the surplus is lost as heat in the crystal

For every electron that is excited, there is also a hole, a positive charge carrier, in the silicon crystal.

The excited electrons in the conduction band are attracted by the n-silicon (see Figure 2.2) and collected on the top of the solar cell. The holes are attracted by the p-silicon and collected in the bottom of the cell. A voltage arises in the solar cell.

If there is a load connected between the front and rear contact, a current arises and the solar cell delivers electrical energy to the load.

The fact that only 1.1 eV of energy can be absorbed by an electron explains why solar cell efficiency is relatively limited. Photons with energy below the band gap give no contribution, and energy over 1.1 eV is lost as heat in the crystal. The theoretical efficiency for a silicon cell is 29%. In laboratory experiments, cells have achieved 25% efficiency.

Other semiconductor materials have different band gaps and thus the theoretical efficiency varies. In tandem cells, solar cells of different materials have been placed on each other. This allows for a greater part of the radiation to be collected. Theoretically, it can achieve 40–45% efficiency. It is solar cells with high transparency, so-called thin-film cells, which are relevant for the technology. They often have low efficiency, which then can be raised to the level of crystalline silicon cells.

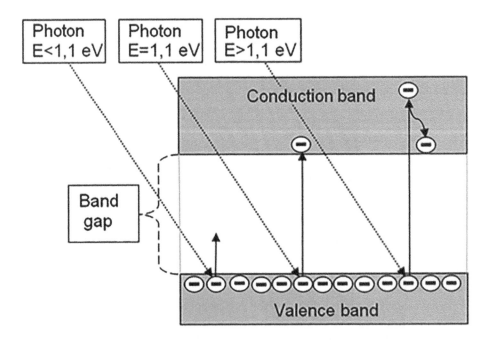

Figure 2.3 Energy diagram showing the electron's energy when it is hit by photons with different wavelengths and energy content.

2.1.3. Electrical properties

No-load voltage of a so-called 12-volt module is about 20 V. It drops when loaded to about 15 V, which is suitable for charging 12-volt batteries. Figure 2.4 illustrates the current-voltage characteristics of a solar module. Curves for two different intensities of solar radiation are shown, 1000 W/m², which is the highest intensity in northern Europe, and 500 W/m².

Characteristic values of a solar module are the voltage at the unloaded module U_{OC} (open circuit voltage) and the short-circuit current, I_{SC}. The voltage U_{OC} is largely independent of radiation intensity. Maximum current I_{SC}, however, is directly dependent on the intensity.

From the diagram we can see that it is not so dangerous to short-circuit a solar cell. The power is very small at short-circuit. But, of course, if many solar modules are connected in series a high voltage and risk for light arcs can occur. If large batteries are connected, the risk is much higher. They can easily produce a few hundred amperes, which can cause material to burn and melt. If we have a PV system for battery charging, it is also important that there is a blocking diode or charge controller/regulator with similar functionality connected; otherwise the battery can be emptied through the solar cells when they are not producing. In Figure 2.4, the load line for an optimal resistive load, R_{OPT} has also been plotted. For best performance, it is important that the load is adjusted to the current production. This load would get maximum power (current multiplied by voltage) at full sunlight. Current and voltage are high simultaneously. However, at half intensity of solar radiation much lower power output would be generated, especially if there is a mismatch with the load.

Charging of batteries is one example of load that is rather well matched with a solar PV module of suitable voltage. With an MPPT controller, the power can be optimized, and in a

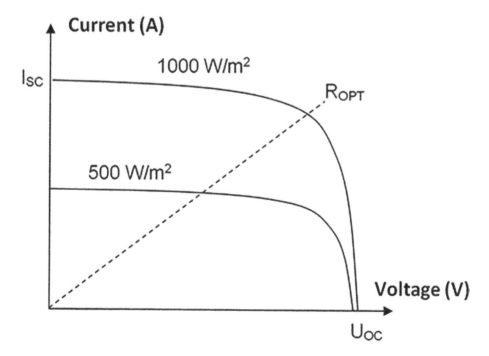

Figure 2.4 The electrical characteristics of a solar PV module with the load line, R_{OPT}, for maximum power output at irradiance (instantaneous solar radiation intensity) 1000 W/m².

battery charging application provides a flexibility to use modules with a very different voltage than the battery charging voltage. MPPT stands for maximum power point tracking, and it is a technology used to find the point at the PV current-voltage characteristics where most power is utilized.

When the power is fed into an electrical grid, a converter/inverter is used that adjusts the voltage and converts it to alternating current. The converter can often also adapt the load so that the output is optimal.

The sun heats up a solar panel. Then characteristics are changed and the efficiency normally drops. Usually, it is important that solar cells are installed in order to provide an air flow under the panels that will cool the cells. An alternative is to combine solar cells with the production of hot water, where the water flow can cool the cells. In a system with batteries, it is important that the batteries are stored in a place with moderate temperatures. Battery life will decrease significantly at high temperatures.

2.1.4. Types of solar cells

Most of today's solar cells are made from crystalline silicon. The cells have a long life and can normally work at least 25 years. Suppliers often give 20-year power warranties. Monocrystalline cells are the type that has the highest efficiency. They have atoms with perfect symmetry in the crystal. In polycrystalline cells, silicon atoms are arranged less symmetrically. The manufacturing is cheaper, but they also have a slightly lower efficiency than monocrystalline cells. Polycrystalline is blue, speckled and monocrystalline smoother dark.

Crystalline silicon cells have achieved efficiencies of more than 23% during laboratory measurements, while the market's modules in a practical setting achieve efficiencies of about 16%. Modules with crystalline cells often have bypass diodes included in the junction box, to avoid the so- called hot-spot phenomena, where one cell that is shaded could have the output of the rest of the module forced through it.

Thin-film cells are another type. The cell consists of a glass plate covered with a thin layer of a photosensitive material. Thin-film cells can be produced at lower cost, but they also have significantly lower efficiency and an uncertainty of lifetime. Thin-film modules often don't require bypass diodes, because of large surface area of their cells (typically in the form of strips) and low cell current.

Currently, three types of thin-film cells are commercially available. They are made of:

- Silicon where the atoms are not arranged in a crystalline structure, known as amorphous silicon (aSi)
- Copper-indium-gallium-diselenide, $CuInGaSe_2$ (CIGS or CIS without gallium)
- Cadmium telluride (CdTe); cadmium is a material that can negatively affect the environment

The active layer of all these is less than 5 microns, compared with the 200 microns in crystalline silicon cells. When they are produced on a large scale, the processes can be automated to a significant extent and the cells can be made in large transparent pieces that give new possibilities in architecture. Thin-film cells of amorphous silicon consist of a thin, dark brown layer on a glass plate.

2.1.5. Adjust a PV system to the load

Often the first step in system design is to try to determine which electrical loads the system should supply. This is especially important for stand-alone systems. The cost of the system will be directly proportional to the amount of energy the system is designed for. Energy from solar cells is still relatively expensive in some applications, so it is important to use the energy efficiently and keep the cost of the system low. Within the EU market today, there are great opportunities to find energy-efficient products thanks to the energy policy tools Ecodesign Directive and Energy Labelling Directive.

The average daily energy demands are obtained from the power in watts of electrical equipment and the daily use in hours. If the system is designed to fulfill the needs in a home, it can be loads such as lighting, coffee makers, microwave ovens, hair dryers, vacuum cleaners, radios, TVs, chargers for phones and computers, power tools and water pumps.

The general approach in selecting electrical loads for cooking and heating should be to consider the opportunity to do it with fuels instead, for example biofuels such as wood, pellets, biogas or propane. Propane can also be used for refrigerators.

For heating, for example of domestic hot water, it is often better to choose solar heating collectors. They have higher efficiency.

When the individual electrical loads have been identified, the power of each load in watts is determined. Nameplates or information from datasheets, manuals, etc., can usually provide this. Loads with variable power demands, such as fans, can be difficult to assess. Here, current or power measurements for the different operation modes can be a possibility. Loads with standby consumption should be avoided, but if it is impossible, loads with very low standby power should be chosen.

The correlation between solar radiation and load over the year is also worth considering. For places with very low solar radiation during winter time, it can be difficult or impossible to power all loads with a stand-alone PV system.

2.1.6. The availability of solar radiation

The amount of energy that a solar module can produce is proportional to the energy of the available solar radiation that hits the surface of the module over a certain period of time. The availability varies from place to place, day to day and time of the year. Most important is that a solar panel is fully exposed to sunlight and not placed in the shadows.

A number of websites provide global information on available solar radiation in exposed areas. One example is www.geni.org. Using words such as "energy" and "renewable energy" and "maps" you can find maps for the energy in the solar radiation in different places in the world. The highest radiation is in desert areas near the tropics.

In Europe, Asia and Africa the website "PVGIS" version 5 [1] is useful. It provides excellent information on the available solar energy resource at different locations. The calculation module is easy to use and also provides a forecast of what PV systems are expected to produce at different slope (also called tilt or inclination) and azimuth angle (that describes the orientation). For example, a two-kW_p (peak power) normally efficient PV system installed with optimum angles provides in Stockholm up to 1750 kWh in a year. The corresponding values in Rome are 2520 kWh, in Tunis 2910 kWh, in the Sahara desert 3800 kWh (best place) and in Cape Town 3020 kWh.

Close to the equator, the variation of the production is small over the year, but in most other countries, the variation is significant. Figure 2.5 shows an example of production for a two-kW$_p$ installation (about 15 square meters) in Tunis in Tunisia. In January the production is 160 kWh and in July it is doubled to 320 kWh.

2.1.7. Selection of PV modules

The power of PV modules (panels) is given in the unit watt peak power, W$_p$. This is the power generated under standardized conditions with an irradiance of 1000 watts per square meter. It is like the sunshine on a sunny day in July in Central Europe. A solar panel of crystalline silicon of one square meter usually delivers about 100–150 W$_p$. The daily performance in watt-hours is calculated by multiplying the power by the time of the power. Although diffuse solar radiation will contribute to the production, the useful power is lower. There are always losses in the system. Examples are the charging and discharging of the battery, the heating due to current in power cords, dust on the panels and so on. In addition to the losses described earlier for the cell/module, normally a system's efficiency is about 60–80% for a wide range of applications.

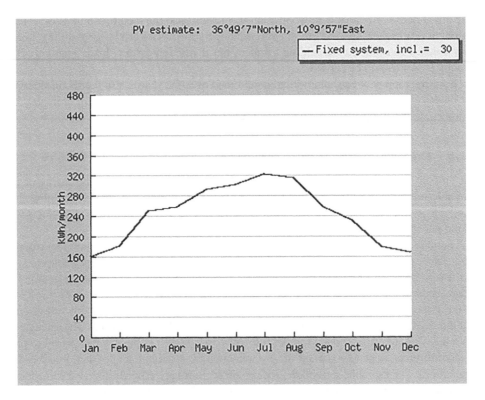

Figure 2.5 Electricity generation in a two-kWp PV installation with 30-degree inclination in Tunis.

Source: Courtesy of PVGIS, JRC, European Commission [1].

Modules for 12-volt systems typically contain 36 cells connected in series, for example, four rows of nine cells in each. Large modules for a 24-volt system have 72 cells.

Most PV panels sold today are of high quality and have good warranties. When purchasing modules, check that the modules are manufactured according to international standards and that manufacturers are certified.

Figure 2.6 shows data for a module of multi-crystalline silicon from the Chinese manufacturer Yingli. They are one of the world's largest manufacturers of solar modules. High efficiency is important if there is limited space for installation, and it also lowers installation costs. In this case, the frames of corrosion-resistant aluminium are tested to withstand high wind and snow loads.

Electrical parameters at standard test conditions (STC)		
Module name		YGE 275
Power output	P_{max}	275 W
Power output tolerances	ΔP_{max}	0 +5 W
Module efficiency	H	16.8%
Voltage at P_{max}	V_{mpp}	31.0 V
Current at P_{max}	I_{mpp}	8.90 A
Open-circuit voltage	V_{oc}	37.9 V
Short-circuit current	I_{sc}	9.35 A
STC: 1000W/m² irradiance, 25°C module temperature.		

Figure 2.6 Data for a multi-crystalline silicon module YGE60 CELL from Chinese manufacturer Yingli Solar. The voltage is suitable for charging two series-connected 12-volt batteries so it can be called a 24-volt panel [2].

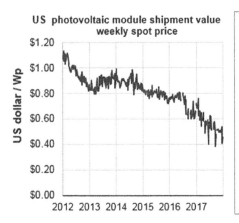

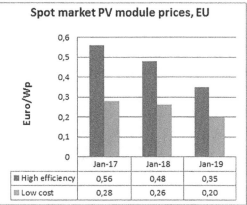

Figure 2.7 PV module price trends 2017/2018. Earlier, 1 €/Wp was a dream target.

Source: Diagram/data from EIA [3] and pvXchange [4].

This product is delivered with a 10-year product warranty and a power warranty: 10 years at 91.2% of the rated power and 25 years at 80.7% of the rated power. If the module is not exposed to extreme conditions it can have a long time span.

In the past several years we have seen a large price reduction for PV modules; see Figure 2.7. The cost of the modules is often approximately 20% of the total cost of a solar energy system. For inverters and batteries, there are no corresponding decreases in prices yet.

2.1.8. Stand-alone applications

Photovoltaics can be used for many applications. Only our fantasy limits ideas for use. Some examples follow:

- Rural electrification: Solar cells can provide electricity to individual buildings, homes, schools, farms and shops. They can, for instance, provide energy for lighting, refrigerators and electronic equipment such as radios, televisions, computers or cell phone chargers.
- Water supply: Pumping of drinking water or for irrigation, pumping of drainage, systems for seawater desalination and water purification.
- Medical and health services: Lighting in rural clinics; refrigeration of blood, medicine and vaccines; sterilization equipment; emergency power at disaster relief sites.
- Communications systems: Telephone, radio and television stations, including aerial equipment; mobile phone towers; telephone boxes; transportation applications. Power for street and railway signals at crossings; emergency telephones at roadways or hiking trails; road and street lighting. Power for lighthouses and buoys at sea. Small-scale supply of various equipment, such as measuring equipment at remote sites, chargers for cell phones and laptop computers, clocks, watches, portable radios, calculators and bike lights.
- Various other applications such as fans for ventilation systems and pumps for solar heating.
- Power supply of holiday homes, caravans, campers and recreational boats; decorative lighting and fountains in the gardens.

Solar panels provide DC power (Fig. 2.8). A DC load with the right voltage demand can be connected directly to the panel. This type of system has weaknesses. It assumes that the operation only needs to be done when there is light. The load must also manage to use the varying power output, which the panel gives according to variations in irradiance.

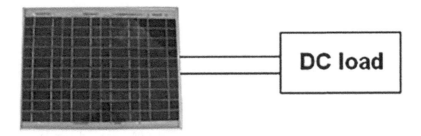

Figure 2.8 The simplest PV solar system.

There are loads that can handle this. Especially good is if you have a function that is related to irradiation.

An example would be an electric fan that ventilates a greenhouse to avoid over-heating. Here, the solar cell is both power source and controls of the function.

The motor of a water pump is another example of a load, which can be powered directly from a PV module or array, if suitable components are used. In developing countries, often sunny, where water supply is a problem, solar-powered pumps can contribute to a solution. Water is pumped during the day up to a reservoir, which can then be tapped when water is needed (Fig. 2.9). Instead of storing the energy, the product is stored and the needs can be satisfied around the clock.

For most loads, we want a function independent of the availability of sunlight. We can achieve this by using battery storage as shown in Figure 2.10a. The solar module charges the batteries and we get electricity from them when we need it.

However, even this system has serious shortcomings associated with batteries. These can be destroyed quickly both by overcharge and deep discharge. Therefore, a charge controller as shown in Figure 2.10b must usually supplement the system. It prevents the battery from overcharging, if use has been lower than the charge during a sunny period.

Deep discharge is also prevented, if the load is automatically disconnected when the battery voltage falls below a certain level, for example, 11.5 V in case of a 12 V battery.

If the disconnection occurs during an exciting stage in a film that you follow on a battery-powered television set, it is tempting to bypass the control unit. The voltage is enough for the TV until the film is over. But then you risk the battery. Many batteries have been destroyed in this way. It is important to understand the controller's role when using a photovoltaic system.

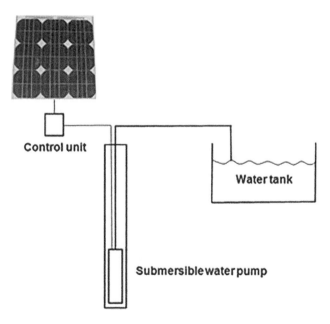

Figure 2.9 Many solar systems for water pumping have been installed, for example in India.

The solar panels and battery must, of course, be designed so that the capacity meets the load requirements. Also be aware that the systems shown here are simplified. For example, fusing is not shown.

Batteries are perhaps the most critical elements in the system. To avoid having a solar system become too expensive, it is of course also important to choose loads that are energy efficient. There is much progress in this area. One of the latest is the LED lamps (light-emitting diode lamps), which have very low consumption and fit well in a DC system. Solar electricity users should also be good energy economists.

The solar systems that we have dealt with so far have had some shortcomings. In a long period without sun, the batteries will run out and we'll have no energy. Furthermore, the systems only provide direct current loads. If we want a system that can meet all our needs and supply us with energy during a period without the sun, we should choose a hybrid system where the solar system has been supplemented by backup power, such as a diesel generator. You can read more about hybrid systems in Chapters 5 and 6.

If we want to use loads with high power consumption, washing machine, wood splitters, etc. the generator can also be run simultaneously, to avoid too high energy output from the battery storage. AC loads can be served, typically by using a grid forming inverter. In some cases an inverter with variable frequency output is an alternative, for example solar pump inverters, which can be based on industrial motor drives including MPPT. There are also other inverters that can power special loads, for example CyboInverter.

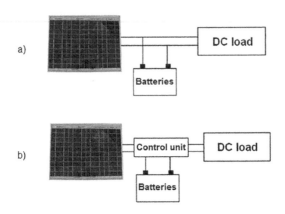

Figure 2.10 Solar systems with batteries.

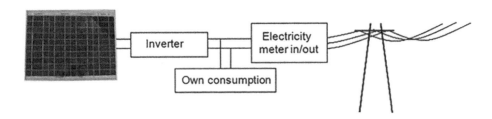

Figure 2.11 System for grid connection.

2.1.9. Grid-connected PV systems

A very large part of today's solar cells are used for grid-connected systems. The installation of such systems is often very simple. Between the PV array and the network, only one grid-controlled (grid-following) inverter, which converts direct current to alternating current, is needed (Fig. 2.11). Easiest is to use PV modules that have a built-in inverter. However, note that it is normally not allowed to plug such an inverter into the nearest outlet, because the inverter should be permanently installed and connected to a dedicated fuse or miniature circuit-breaker (MCB). Also, to connect one's own production unit to the grid requires an agreement with the network owner and sometimes additional equipment to measure the delivered current. The equipment is often expensive and there are usually certain installation costs regardless of size, so for small plants the income may be too low. In some countries, a significantly higher price is paid for the delivered energy. In Germany, power from photovoltaic was earlier paid up to 0.4 €/kWh, including subsidies.

2.1.10. Photovoltaic in developing countries

Over one billion people, a significant part of the world population, lack electricity. Another large number only have access to electricity for limited hours. Electricity grids are extended, but it is costly, and often there is a lack of a sufficient customer base to make the grid profitable. In many villages, one can only obtain a very small sale of electric energy. In many areas, distances can be long between villages, so the power lines are long and require a high investment cost. Besides, the infrastructure to operate power networks and production facilities and to ensure fuel supplies is often lacking. Decentralized small systems based on solar cells can be the solution in many places.

Calculations have shown that with one solar panel per person, a number of basic functions can be obtained, such as water supply, disinfection of water, home and street lighting, television, refrigerator in the village shop and more. It would radically change the lives of many people.

In one project, SPOTS (The Solar Power Technology Support) was implemented in the Philippines' Mindanao region; 114 villages were electrified with solar cells. The project was funded by Spanish development assistance funds.

The project included:

- 15,000 home-lighting appliances
- 69 solar-powered irrigation systems
- 97 solar-powered pumps to supply drinking water
- Basic electrification of 147 schools, 148 village centers and 27 health clinics
- 293 public lighting points
- 79 units for AC power to workshops in villages for gainful employment

The project is expected to increase productivity in agriculture and industries and increase household incomes and improve living conditions for 400,000 people in the most isolated villages in Mindanao.

Africa has good access to renewable energy resources. Hydropower has long been important and there are prerequisites for further expansion. But the continent also has good conditions for more solar power. The cost reduction for solar cells in recent years has meant that these provide a quick and cost-effective way of giving more people access to energy. Still, approximately 600 million Africans today do not have access to electricity.

According to a report from The International Renewable Energy Agency (IRENA) [5], the cost of solar power in Africa has decreased rapidly during the last years. Another advantage is the short project lead times and rapid deployment of solar PV, compared to many other generating options.

The report shows that mini-grids with solar cells or off-grid solar home systems provide a high quality of energy services at the same or lower cost than the alternatives. Independent solar PV mini-grids have been installed in Africa at a cost of US$ 1.90 per watt for systems over 200 kW. In suitable conditions, small off-grid electricity systems can give households their annual electricity requirement at a cost of US$ 56 per year. IRENA estimates that Africa with a more active policy could get more than 70 GW PV with solar power by 2030.

A new technology for lighting, energy-efficient LED lamps, is now available at lower prices than before and are often used in solar systems. That improves the possibilities of arranging additional lighting with solar cells. LED lamps consume significantly less energy than other light sources, so you can get good lighting with smaller solar panels. They also have an extremely long lifespan, up to 100,000 hours vs. 1000 hours for conventional bulbs. It is particularly important in developing countries where the parts supply system often is badly developed.

Kerosene and other fuel-based lighting account for a considerable portion of the world's total cost for lighting but provide only a limited part of the practical exchange of lighting. The operational cost of a kerosene lamp can be 150 times greater than for an LED-system powered by solar energy, but the new system requires new investment.

2.1.11. Batteries in solar systems

A PV module will produce electricity when the sun shines on the solar cells. In an off-grid system, energy storage is necessary to provide solar electricity at night or on cloudy days. Battery systems typically are dimensioned to provide electricity for some cloudy days. Even grid-connected systems can include, for example, batteries as a backup during a power failure. It may be useful to keep critical electrical loads running until power is restored. Examples of loads that may need backup are equipment for telephones, computers and lighting.

Ordinary starting batteries for cars are usually not suitable. They are designed to provide a high power for short periods, but can't manage many large load cycles.

Batteries used in solar electric systems have to cope with many large charges and discharges. Suitable batteries are often called solar batteries. Small batteries are sometimes integrated with a solar panel or solar powered lamp. To read more about batteries, see Chapter 3.

2.1.12. Risks of accidents and injures

A PV array with a so-called extra-low voltage, for example max 75 V DC according to EU directive, gives a minimal risk of electrical accidents. Such voltage is likely too low to cause severe electric shock or cause damages by arcing or melting material.

However, when several modules are connected in a series, for example to a grid-connected inverter, this will increase the voltage to dangerous levels and regulations for voltage higher than 75 volts must be applied. There should be devices that can separate equipment safely from voltages both on the DC side and AC side. Note that connectors should not be opened under load. To avoid working when voltage is present, it can be useful to first cover the PV array or to work at night.

For a stand-alone facility, it is more likely batteries can be a risk factor, even if the voltage is low. Shorting a battery can provide high currents and arcing, that can cause burn damages and risk for fire. When working at a power plant it is important that there are devices for disconnection of dangerous voltage from all sources.

The glass covering of a solar module is usually not any stronger than an ordinary pane of glass, so it is easy to damage it, accidentally or through sabotage, with blows or stone throwing. In some places, it may be wise to place panels where they are not easy to discover, in order to reduce the risk of sabotage or theft. Heavy snow can in some cases also cause damage.

2.1.13. Large-scale power plants

Most installations of solar cells have been done on a small scale on individual houses. But in recent years, interest has increased for the use of solar energy for large plants, where thousands of modules are installed together in a large power plant.

As an example, "Solparken" in Västerås, Sweden, went into operation in 2014 (Fig. 2.12). It consists of 92 solar tracking units with 36 solar modules in each. The total power is 1 MW and the park will produce 1.2 GWh per year. The arrays rotate and tilt towards the sun continuously. Thanks to this, the relative production increases compared with parks with non-moving units, but the investment costs are also increased considerably.

There are single axis and double axis tracking systems. Single axis systems give 30–35% more production than fixed; in double axis systems the production increases another 5–10%. The figures can differ depending on the location on Earth.

Figure 2.12 The arrays placed on trackers in the photovoltaic power plant in Västerås, Sweden, follow the sun's movement.

Source: Lasse Fredriksson [6].

2.1.14. Solar cell history

The French physicist Edmund Becquerel, then 19 years old, discovered in 1839 what became known as "the photovoltaic effect" [7]. Becquerel observed that when two illuminated metal electrodes immersed in a semi-conducting electrolyte, a weak electrical current occurred. What he discovered was that the sunlight was energy, which by means of semiconductors could be converted into electricity. Becquerel's observation was considered very interesting, but it found no practical application.

Bell Telephone Laboratories in New Jersey began to develop today's solar cells in the 1940s. It was surprisingly found that the silicon wafers doped with certain impurities were very sensitive to light. In 1954, Bell had made a solar cell with an efficiency of 4–6%. The following year the first solar cells were released into the commercial market. In the following years the company managed to improve the efficiency to 8–9%. A breakthrough for the use of solar cells came in 1958, when the first satellite with solar cells, Vanguard I, was launched. The solar cells worked so well that the space industry believed it was an effective energy source for spacecraft and the cells since then have been a part of space programs. But the solar panels were expensive and therefore the market for them was very limited. Development continued and terrestial use received a significant boost after the first oil crisis in 1973 [8]. After a slump during Ronald Reagan's presidency in the US, the research took on a new impetus after the Chernobyl nuclear plant accident.

Today there is strong agreement on the need to invest in this sustainable energy source. At the end of 2018, more than 500 GW PV cells were in use worldwide. Most of the installations were grid-connected.

2.1.15. Concentrating solar power

Even solar heat can be used to produce electricity. If solar radiation is concentrated by reflection, sufficiently high temperature to produce steam is achieved. Solar energy can be concentrated by parabolic mirrors as in Figure 2.13. Solar radiation is focused on a tube in the middle of the mirror. The mirrors are moved by a control system, so that they can get the best angle to the sun. Oil in the central tube is heated to high temperatures through a heat exchanger; water can be boiled into steam, which may then drive a steam turbine and generator in a conventional manner. Figure 2.13 was taken at the power plant SEGS (Solar Electric Generating Station) in the Mojave Desert in California. SEGS consists of nine power plants with outputs ranging from 14 to 80 MW, totalling 354 MW.

There are also small concentrating solar power (CSP) plants using, for example, Stirling engines or micro turbines to drive a generator.

One drawback with many solar power stations is that they can't produce electricity on cloudy days or at night. Some CSP plants are therefore of hybrid performance. First, the power plants have an energy storage that can handle short interruptions of solar radiation, and they have an additional fuel-driven (usually natural gas) unit for generating steam, capable of operation at longer periods without sun radiation.

Another form of thermal solar power plant is called central tower power. Solar radiation is concentrated at the top of a tower in the middle of a field. Thousands of mirrors, continuously rotated and tilted, reflect the radiation. At the top of the tower, mineral oil is heated to a very high temperature. The oil then can be used to generate the steam needed to operate a steam turbine and then generate electrical power in the usual way.

Figure 2.13 Concentrating mirrors in one of the world's largest concentrating solar power plants in California.

Source: DOE/NREL.

2.2. Wind power

2.2.1. Introduction

The wind blows all over the world. Everything that moves carries kinetic energy and so does the air in the wind. Wind energy can be described as indirect solar energy. About 1% of the energy in the solar radiation that hits the earth becomes converted to wind energy. With a modern wind turbine, up to 50% (theoretical value 59%) of the energy in the wind can be converted to electric energy. However, the wind resource is not evenly distributed in space and time.

A challenge with wind energy is that it varies as the wind does. Therefore, it must usually be supplemented by other energy sources. In many large-scale power systems in the world, hydropower with its dams is a good complement. The interconnection of grids and trade with electrical power over borders is also helpful. In both large and small power systems, energy management and storage systems, including new batteries, can also improve the situation, so that wind can become an even more important energy source in future energy systems.

In countries such as Sweden, an advantage of wind power is that the average energy content of the wind is highest during the dark months of the year, when we need energy the most (Fig. 2.14).

Figure 2.14 Five 2.3-MW turbines operate in the harbor area in Falkenberg, Sweden. The municipality intends to become self-sufficient on electricity.

Source: Margareta Gunnarsson.

2.2.2. Wind energy history

Humankind has harnessed the wind's energy for thousands of years, in many different parts of the world. Vertical-axis windmills are known from old China and Persia. For more than 800 years the horizontal-axis European windmill has been used for a variety of duties, but mostly for grinding grain and pumping water. The windmill's modern equivalent – the wind turbine – is mainly used to generate electricity.

James Blyth, a Scottish engineer and university teacher, began experimenting with windmill construction and in July 1887 he had constructed a small windmill for supplying electric light by means of a battery in a garden. In the US and Denmark, wind machines for electricity generation were also constructed early.

The first wind turbine that generated electricity for the public grid was constructed in Gedser in Denmark in 1956. It operated well for 11 years. But at that time, with very cheap oil, it was more profitable to generate electricity with fossil fuels, and the turbine was taken out of operation. During the oil crisis in 1973, the price of oil increased and interest in wind power awakened. At the beginning of the 1980s, several manufacturers in Denmark had started to produce wind turbines. Many of these turbines were exported to California, and the new era of wind power began.

Today wind turbines are available in all sizes up to about 6 MW rated capacity and are considered a mature technology. Small wind turbines (usually defined as <100 kW-rated power) are often used in locations far from the power grid to generate electricity for local needs or to pump water, etc. Medium (100–1000 kW) and large (>1 MW) wind turbines are usually made for grid connection.

2.2.3.   Components of a wind turbine

Wind turbines are mounted on a *tower* or guyed *mast* to capture more energy. At 20 meters and higher above ground, the wind is stronger and less turbulent than below. Large wind turbines today have towers that are 100–150 meters high. In Gaildorf, near Stuttgart in Germany, a wind turbine has been constructed with a hub height of 178 meters – the tallest in the world [9]. It is combined with a pumped hydropower station.

Turbines catch the wind's energy with the *rotor blades* (Fig. 2.15). Usually, three blades are mounted to a hub to form a rotor. Sometimes more than three blades are used on small turbines. Reasons for that can be to have a higher torque at low wind speeds or to lower the rotational speed of the blades and so that the turbine is quieter. Two-bladed rotors can be of interest, for example, for off-shore wind turbines, where installation and access by helicopter may be simplified.

The rotor is connected to a drive train in the *nacelle*. It often consists of a *gearbox* and a *generator*. The gearbox increases the rotational speed to be suitable to the generator.

Small wind turbines and some big turbines with a multi-pole generator have no gearbox. They are called direct drive turbines. Some advantages of direct drive are that the gearbox is an expensive component requiring maintenance. But direct drive generators are much heavier, especially for large turbines, so it is not easily determined which of the concepts will be more common in the future.

The wind turbine generator converts the rotational energy to electrical energy. The generators have to work with a fluctuating power source. Grid-connected generators normally produce alternating current (AC) with the voltage and frequency of the grid, for example 230/400 V, 50 Hz (or higher voltage for large generators).

Generators in small wind turbines are often supplemented by a rectifier to produce a voltage that is suitable to charge batteries, for example 12- or 24-volt direct current (DC).

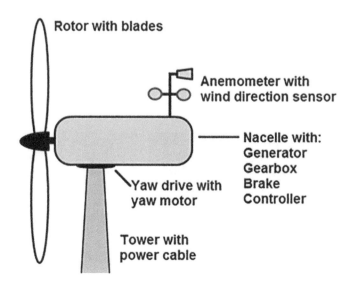

Figure 2.15 Basic parts of a typical wind turbine.

Wind turbines with variable rotor speeds produce a current, which must usually be converted to the right voltage and frequency by a frequency converter, often located at the bottom of the tower.

Most wind turbines are designed for the rotor disc to be upwind from the tower, but some operate with the rotor disc downwind.

The yaw drive with a yaw motor keeps the wind turbine in the right direction to catch the energy of the wind efficiently. The controller receives necessary signals from the wind direction sensor at the nacelle to control the yawing.

Small wind turbines often have a tail or downwind design that keeps the turbine in the right direction. The tail can also be part of a furling system, where the rotor disc is turned away from the direction of strong winds, without any active control system.

2.2.4. *Fundamentals of electricity generation*

The blades act like airplane wings. When the wind blows, low pressure is formed on the downwind side of the blade and high pressure on the upwind side. The pressure difference then pulls the blade, causing the rotor to turn. The turning shaft spins the generator, which converts the rotational energy to electrical energy.

Wind turbines are classified as horizontal-axis (HAWT) or vertical-axis (VAWT). Conventional wind turbines are of the horizontal-axis type. The modern, highly efficient horizontal-axis wind turbines for electricity production have rotors with typically three slender blades, designed for the speed of the blade tips to be much higher than the wind speed (high tip-speed ratio). This design is optimal for electricity production.

Mechanical wind pumps normally use a multi-blade rotor with large blade area, which creates a high torque at low wind speeds. These wind pumps are optimized to pump water as much of the time as possible. They have a low tip-speed ratio, which also makes them quieter.

The size of a wind turbine is best described by the swept area of the rotor (or by the rotor diameter, in case of a circular swept area), because that tells us how much of the wind stream the rotor can intercept. The height of the wind turbines also has a significant impact on produced power. It is usually specified as the hub height, which is the height of the center of the rotor above ground.

Various types of wind machines are illustrated in Figure 2.16. Totally different concepts, for example based on kites, have also been proposed.

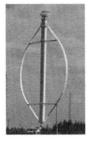

Figure 2.16 Some types of wind machines that have been built through the years: *left to right*, an old windmill, a two-bladed Darrieus turbine, a common three-bladed turbine, an early two-bladed megawatt prototype and a vertical-axis H-turbine.

2.2.5. The wind

At some areas on the ground the air becomes more heated. Warm air is lighter than cold and it will begin to rise. A low-pressure area forms. Air in surrounding areas will move into the low-pressure area, creating wind. Places that will be heated more are, for example, areas close to the equator, land vs. sea areas and mountain slopes that face the equator. The tradewind, the monsoon, the sea breeze, the land breeze, mountain winds – all are driven by differences in solar heating.

As we all know, the wind differs from time to time. Figure 2.17 demonstrates how energy generated in a wind turbine can vary during 15 minutes. During this time period the power varies between 0 and 200 kW. The power is proportional to the third power of the wind speed. Wind turbines with varying rotational speeds can absorb a part of the variation by changing the rotational speed, so they give a little smoothed power output.

The easiest way to describe the wind at a certain place is the average wind speed, often given over a year. For wind data, the height is also important to know. Meteorological data for wind often refers to 10 meters above ground. Typically the average wind speed is between 4 and 9 meters per second (m/s), in open areas.

A better way to describe the wind is by using the frequency of different wind speeds. This is often simplified by using a Weibull distribution. An example is given in Figure 2.18. The distribution is described by the two parameters: the scale parameter, A, and the shape parameter, k. If the shape parameter k is exactly 2, the distribution also is called Rayleigh distribution.

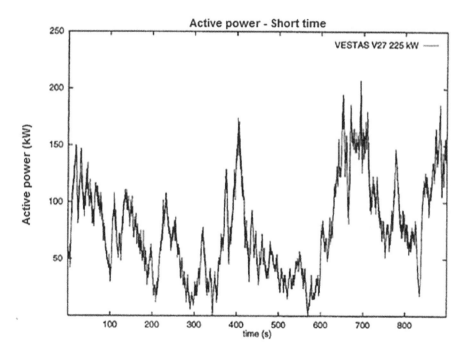

Figure 2.17 Variations in the energy in the wind speed that a wind turbine collected, measured for 15 minutes.

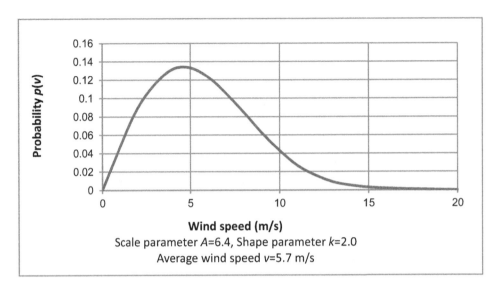

Figure 2.18 Example of the Weibull distribution, which shows the probability for the wind to have a certain speed.

Table 2.1 Examples of roughness classes and energy indexes at different landscape types

Roughness class	Energy index (%)	Landscape type
0	100	Sea and lake, water surface
1	52	Open agricultural area without fences and hedgerows and very scattered buildings; only softly rounded hills
2	39	Agricultural land with some houses and 8-meter-tall, sheltering hedgerows with a distance of approximately 500 meters
3	24	Villages, small towns, agricultural land with many or tall, sheltering hedgerows, forests and very rough and uneven terrain
4	13	Very large cities with tall buildings and skyscrapers; high and dense forests

2.2.5.1. Wind shear and roughness classes

Wind speed is influenced by the surface of the ground. Friction against the surface of the earth decreases the speed. The speed is also affected by various obstacles such as bushes, hedges, large trees, forests and buildings. Also the terrain's appearance with slopes, hills and mountains affects the wind speed.

In the wind power sector, the impact of the terrain and the surface area on wind speed is often described with roughness classes. Examples are shown in Table 2.1.

The wind speed also varies with the height above the ground. This is called *wind shear*. Three examples are shown in Figure 2.19. The higher the roughness class is, the lower the wind speed close to the ground will be.

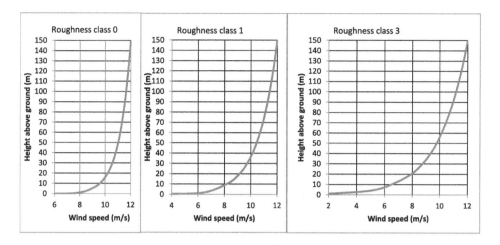

Figure 2.19 Examples of wind shear – the variation of wind speed with height. The figure shows the wind speed at different heights at areas with roughness classes 0, 1 and 3, if the speed is 12 m/s at a height of 150 meters.

This indicates that the tower of a wind turbine ought to be much higher at areas with a higher roughness class. And even the opposite is valid, at sea with roughness class 0, the towers can be relatively low.

At a smooth land area, for example an ordinary farm landscape, the variation in the wind speed will give about 1% higher production with a tower that is 1 meter higher, in the range of 20–40 meters above ground level.

2.2.5.2. Wind measurement

To determine where to place wind turbines, more detailed wind data can be obtained by wind speed measurements at the site.

Wind speed is usually measured with a cup anemometer. It is often equipped with a wind vane for the wind direction. The best place for an anemometer is on the top of a mast at the proposed turbine's hub height.

Best of all is if long-term wind speed measurements of high quality are available. It is recommended to measure at least one full year. Wind data measured on the site can then be correlated with long wind measurements from other sites in the region.

Wind speed is usually measured as 10-minute averages: minimum, maximum and standard deviation. The standard deviation is a measure of the turbulence. Data on wind speed and direction are normally recorded in a data logger for later long-term evaluation.

A first step to mapping the wind resources can be with SODAR (SOnic Detection And Ranging) measurement (Fig. 2.20). SODAR is a type of acoustic radar. Sound pulses are sent out and reflected sound is received by a device at the ground. With SODAR, wind speed and turbulence intensity can be measured, for example from 20 to 150 meters in steps of 5 meters. Data are transferred, for example via the mobile phone network, to a computer. SODAR measurement for three months gives a good indication whether the site is suitable

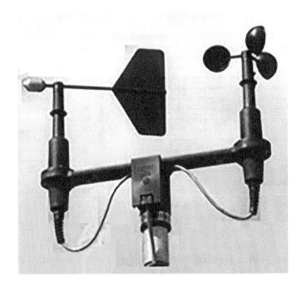

Figure 2.20 Equipment for wind speed measurement: a cup anemometer with a wind vane and a trailer with equipment for SODAR measurement.

Source: Vindtek AB [10].

for wind power. The equipment is simple to establish. Only a site to park a car trailer is needed. With SODAR, expensive measurement installation and planning permission can be avoided. LIDAR (LIght Detection And Ranging) is a technology that can be used for similar applications, but works with laser light instead of sound. Where indications justify, more careful wind measurements can be made, typically with a wind mast with anemometers and wind vanes on several levels.

For a small wind turbine, the procedure just described often cannot be justified economically. If other wind turbines are in operation nearby, their production data are also valuable for a wind resource assessment.

An alternative can also be to install a very small and cheap wind turbine and measure production for a time, especially if the turbine has a known power curve. The production from that can be a good base to judge whether a more expensive wind turbine or additional turbines in the area can be profitable.

2.2.5.3.   Energy in the wind

The energy of the wind is kinetic energy. This is given by the equation:

$$W_{kin} = \tfrac{1}{2}m \cdot v^2$$

(2.1)

where: W_{kin} = kinetic energy, m = mass, v = speed

The mass that passes the swept area of a wind turbine depends on the air density, the area, the wind speed and the time:

$$m = \rho \cdot A \cdot v_{wind} \cdot t$$

(2.2)

where: ρ = density, A = swept area, v_{wind} = wind speed, t = time

Air density is dependent of air pressure and temperature (and to some extent on humidity). It is higher in low temperatures and high air pressure. In wind power calculations a standard figure of $\rho = 1.225$ kg/m^3 for the air density is commonly used. That is the density at normal atmospheric pressure at sea level (1013 hPa) and 15° C. For many sites these figures give a satisfactory accuracy to the calculation.

By inserting equation (2.2) into (2.1), we get the energy of the air mass that passes the swept area:

$$W_{wind} = \tfrac{1}{2} \cdot \rho \cdot A \cdot v_{wind}^{3} \cdot t$$

(2.3)

The power in the wind that passes the swept area is calculated by dividing with time:

$$P_{wind} = \tfrac{1}{2} \cdot \rho \cdot A \cdot v_{wind}^{3}$$

(2.4)

This is an important equation. It shows that the power is proportional to the cube of the wind speed. Even a small change in the wind speed gives a big change in the power. Therefore, it is important to find sites with good winds and reach high enough with a sufficient swept area to capture it.

2.2.5.4.   The cube factor

If you know the mean or average wind speed and want to calculate the average energy in the wind, you can use the cube factor, k_3. The formula becomes:

$$P_{wind,av} = \tfrac{1}{2} \cdot k_3 \cdot \rho \cdot A \cdot v_{wind,av}^{3}$$

(2.5)

where: $P_{wind,av}$ = average power in the wind, $v_{wind,av}$ = average wind speed, k_3 = the cube factor

The cube factor, sometimes called energy pattern factor (EPF), is dependent on the wind speed distribution. If the shape parameter in the Weibull distribution is 2, then the cube factor k_3 =1.9. Thus, the average power at the average wind speed of 6 m/s is, with this wind distribution, nearly double compared with the constant wind speed 6 m/s. The physical cause of the cube factor is that at winds over the average wind speed, the contribution of power is much larger than at wind speeds below the average. In approximate calculations you can calculate with the cube factor 1.9. At sites with significant variations in the wind this can underestimate the energy and if the wind is steady, for example the trade winds in the Caribbean, the energy can be overestimated. To do careful calculations you must know the wind distribution, for example the Weibull parameters.

2.2.5.5. Betz's law

It is impossible to convert all the energy in wind with a wind turbine. If it were possible, the wind speed behind the wind turbine would be zero. The German physicist Albert Betz showed in his book *Wind Energie* in 1917 that it is only possible to convert 16/27, or 59%, of the wind's energy to mechanical energy in a wind turbine.

2.2.6. Power of the wind turbine

A completely ideal wind turbine with no losses in bearings, gearbox and generator should produce 59% of the power in the wind. Real wind turbines give less. The most efficient wind turbines can give up to about 52% of the energy in the wind at the most efficient point of operation. A general expression for the efficiency is the power coefficient, C_p:

$$C_p = P_{turbine} / P_{wind} \qquad (2.6)$$

where: $P_{turbine}$ = power output of a wind turbine, P_{wind} = power in the wind

If you know the power coefficient for a certain wind turbine, you can calculate the power with the formula:

$$P_{turbine} = \tfrac{1}{2} \cdot C_p \cdot \rho \cdot A \cdot v_{wind}^3 \qquad (2.7)$$

The power coefficient varies for different wind speeds and different rotational speeds of the turbine. It also varies for different wind turbines and can be reduced over time, for example because of dirt accumulating on the leading edge of the blades. In different turbines and operational states the power coefficient can vary between 0 and 0.5. An average value for a large wind turbine can be $C_{p,av,large}$ = 0.4. Small wind turbines are less efficient than large and often don't operate at their optimum point, so the average value can be, for example, $C_{p,av,small}$ = 0.25.

Figure 2.21 shows how much power is accessible in the wind at different wind speeds and how much an ideal and a modern wind turbine can produce. At 10 m/s the energy in the wind amounts to more than 600 watts per square meter. An ideal wind turbine produces 400 watts and a real modern turbine 300 watts. A small wind turbine would typically give less.

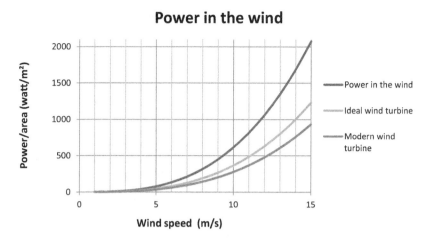

Figure 2.21 Wind power and how large a share an ideal and a modern wind turbine can catch.

2.2.6.1. The power curve

The power curve shows the power a wind turbine will give at different wind speeds. Normally, the power curve is measured according to an IEC standard, at least for every large wind turbine on the market. You can get the power curve from the supplier or sometimes from independent sources, which have performed their own measurements. However, be aware that some power curves are published that are based only on theoretical calculations, estimates or poor measurements. Ask how the power curve was established, if it is not clearly stated.

Nothing tells you more about a wind turbine's potential than the power curve. Professional procurement of wind turbines is therefore often based on the power curve, and specifies penalties if the wind turbine does not perform according to the power curve provided by the manufacturer.

Figure 2.22 shows the supplier's data for a small (3-meter-diameter, 3 kW) wind turbine. It is based on a Chinese turbine sold, for example, in Europe under different brands. The wind turbine can be used both in DC power systems or in an AC system with a converter. The yearly generation at a site with moderate wind resources is about 4000 kWh, according to the data provided by the supplier. That is a normal domestic electricity need in a European household.

The power curve according to the supplier is shown in Figure 2.23. The wind turbine starts to produce energy at the cut-in wind speed, 2.5 m/s. At 13 m/s this turbine gives the maximal power of approximately 3500 watts.

The problem is to know if the data are true or not. We are not aware of any independent measurement that verifies the claims of the supplier. If the data is true, then the turbine would be amazingly efficient. The turbine would produce much more per square meter swept area than any of the small wind turbines that have been properly tested according to standards for a consumer label – see Table 2.1. The rated power is also relatively high compared with the swept area, which may also cause a buyer to think that the production will be high.

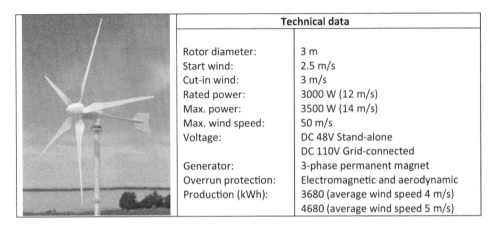

Technical data	
Rotor diameter:	3 m
Start wind:	2.5 m/s
Cut-in wind:	3 m/s
Rated power:	3000 W (12 m/s)
Max. power:	3500 W (14 m/s)
Max. wind speed:	50 m/s
Voltage:	DC 48V Stand-alone
	DC 110V Grid-connected
Generator:	3-phase permanent magnet
Overrun protection:	Electromagnetic and aerodynamic
Production (kWh):	3680 (average wind speed 4 m/s)
	4680 (average wind speed 5 m/s)

Figure 2.22 Supplier's data for a small wind turbine, Windstar 3000.

Source: Windforce [11].

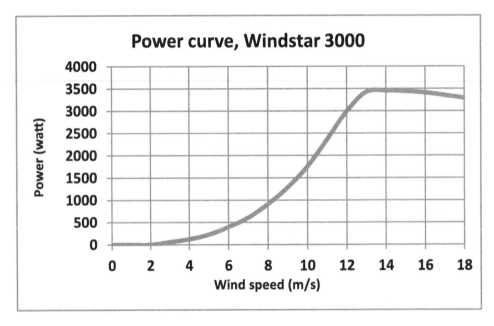

Figure 2.23 The power curve of a 3-meter-diameter wind turbine, Windstar 3000, according to the supplier Windforce [11].

At higher wind speeds the power must be limited to avoid overload of the mechanical parts and generator. The power control can be done in different ways. Large wind turbines often use pitch control. The pitch angle of the blades is then changed so that they catch less of the power in the wind. Another way is stall control, an aerodynamic method. In that case, the blades have a fixed pitch angle and are designed in such a way that air whirls arise behind the blades at wind speeds over the nominal and limit the production.

Small wind turbines are often controlled by furling out of wind, vertically or horizontally. This can be done by a special design of the tail vane. At wind speeds over the nominal, the rotor is turned out of the wind so the area that faces the wind becomes smaller and the power decreases. Furling can cause a drastic decline in the power curve at high wind speeds, but that is usually not anything to worry about, because high wind speeds are not so common (and e.g. in battery-charging applications, the batteries would probably be full anyway during a high-wind period).

Large wind turbines usually have a cut-out wind speed of approximately 25 m/s, when they automatically shut down. That may at first appear like a significant drawback, but it is normally not, because higher wind speeds are so rare. When looking at a power curve, the most important part is normally up to approximately 10 m/s, because higher wind speeds are not as frequent on most locations.

Most power curves are published for the standard air density 1.225 kg/m³, and wind speed in a power curve normally refers to hub height of the wind turbine. Note that power curves are usually based on measurement data, where all wind speed and power measurements have been averaged, typically over 1 minute, 10 minutes or 1 hour. Therefore, very short power peaks and dips are not shown by the power curve. Also, be aware that long-term averages, such as annual average wind speed, should never be used with a power curve.

2.2.7. Calculating energy production

For simplicity, we here assume that the wind turbine will have 100% availability, which means that no production will be lost as a result of repairs, etc.

Equations (2.5) and (2.6) give an opportunity to calculate the energy production. If you combine them you get the average power of the wind turbine:

$$P_{turbine,av} = \tfrac{1}{2} \cdot C_{p,av} \cdot k_3 \cdot \rho \cdot A \cdot v^3_{wind,av}$$

(2.8)

By multiplying the average power with the hours of the year you get the annual energy output:

$$W_{year} = 8760 \cdot P_{turbine,av}$$

(2.9)

Example 1: The wind turbine in Figure 2.22 has a diameter of 3 meters. At the site in Figure 2.18 the average wind speed is v_{av} = 5.7 m/s. By using air density ρ = 1.23 kg/m², cube factor k_3 = 1.9 and average power coefficient $C_{p,av,small}$ = 0.25 we get the average power and the annual energy output of the wind turbine as shown here. Note that in this case the estimation with this method gives much lower results than the supplier's data.

$$A = \pi \cdot 3^2 / 4 = 7.1 \text{m}^2, P_{turbine, av} = \tfrac{1}{2} \cdot 0.25 \cdot 1.9 \cdot 1.23 \cdot 7.1 \cdot 5.7^3 = 384 \text{W}$$

$$W_{year} = 8760 \cdot 384 \text{ Wh} = 3364 \text{kWh}$$

A more accurate method to calculate the production is to us the wind distribution (e.g. the Weibull curve for the wind) and the power curve for the wind turbine. Computer programs use this method, for example the Danish program *Windpro* and a calculator on a website (see later).

Example 2: The power curve can be expressed in figures instead of a graph. We get for the power curve that shown in Figure 2.23:

Wind speed (m/s):	3	4	5	6	7	8	9	10	11	12	13	14	15	16	18
Power (kW):		0	0.1	0.2	0.4	0.6	0.9	1.3	1.8	2.4	3.0	3,4	3.5	3.4	3.4	3.3

On the website of the Danish Wind Industry Association, www.windpower.org, you can find some calculators. Click on "English," "Wind in Denmark," "Education," "Windpower Wiki," "Energy output" and "The power calculator". Feed the calculator with the figures for the power curve and the wind data (Weibull parameters) from Figure 2.18. Do not use a lower hub height than 18 meters. Click on "calculate" and you get the answer: W_{year} = 5081 kWh. This calculation is more accurate than the one in example 1, if the power curve is correct.

To get a preliminary rough estimate of the production of a particular horizontal-axis wind turbine with a rotor diameter d (in meter) at a certain average wind speed, use equation (2.10):

$$W_{year} = 2 \cdot d^2 \cdot v_{wind,av}^3$$

(2.10)

Example 3. With a diameter of 3 meters and an average wind speed of 5.7 m/s the production can be estimated with equation (2.10) to:

$$W_{year} = 2 \cdot 3^2 \cdot 5.7^3 = 3333 \text{ kWh}$$

This calculation is easiest to do and can be used for a first estimation of which size of wind turbine to choose. You don't need the power, only the diameter of the wind turbine with this equation. The diameter (or swept area) is the most important size data of a wind turbine.

2.2.8. Large grid-connected wind turbines

Grid-connected wind turbines at utility scale in addition to solar PVs have become the fastest growing electricity generation methods. The installed power has grown almost fivefold in 11 years. Figure 2.24 shows the development. A common size of the large wind turbines erected

Total global wind power installations

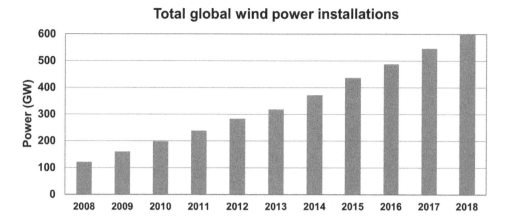

Figure 2.24 Total installed wind power worldwide.

Source: Statistics from: WWEA [12].

Vestas V90-2.0 MW, Technical Data	
Rated power	2/2.2 MW
Rotor diameter	90 m
Hub height	80/95/105 m
Cut-in wind speed	4 m/s
Rated wind speed	13.5 m/s
Cut-out wind speed	25 m/s
Generator type	Doubly fed induction generator, slip rings
Swept area	6362 m²
Rotational speed	Max. 17.6 rpm
Regulation	Pitch-regulated with variable speed

Figure 2.25 Data for a grid-connected large wind turbine.

Source: Vestas [13].

today is 2–4.5 MW. At a medium site on-shore the production can be about six million kWh per year for a 2-MW turbine. An off-shore site can give 50% more.

Figure 2.25 provides the technical data of a typical large wind turbine, the Vestas V90 2-MW wind turbine. The company has delivered almost 8000 of these wind turbines all over the world. It is pitch regulated, which was described earlier. This wind turbine has a double-fed induction generator, which gives a limited variation of turbine rotational speed.

2.2.9. *Small wind turbines*

Wind turbines with power up to 100 kW are usually considered small wind turbines. According to estimations from the World Wind Energy Association (WWEA), a total of 990,000 small wind turbines with a total power of 945 MW had been installed at the end of 2015.

Average rated power of the turbines was 958 watts. This number has increased by about 8% a year over the past five years, so the trend is towards larger machines.

The increase in total installed power during the year was 118 MW. According to the forecast from the WWEA, the rate of increase will continue over the next five years, so 2000 MW should be reached around the year 2020. However, the report does not seem to include decommissioning of turbines.

According to the WWEA, about 330 manufacturers of small wind turbines were identified in 2011. Half of the manufacturers are in five countries: Canada, China, Germany, UK and USA. China and USA were reported to be the largest markets for small wind (Fig. 2.26).

The traditional HAWT dominates the market. Only 14 of the manufacturers solely produce VAWTs. VAWTs suffer from lower efficiency, but VAWT proponents may see niche markets (such as very turbulent or noise-sensitive sites) as an opportunity. Figure 2.27 is of a VAWT that is developed by Uppsala University.

Figure 2.28 illustrates a simplified electrical schematic of a typical small, stand-alone wind power system, based on battery storage. It can be used for rural energy applications. Wind turbines for this application are typically available with power ratings up to 10 kilowatts and a diameter up to 7 meters, but also a few larger types of wind turbines can be found that are suitable for use without connection to the national grid.

Despite a market trend of grid-connected systems with greater capacity, so is off-grid use important primarily in remote areas. Some applications are electrification in areas lacking a power grid and supply to telecommunications facilities. Small wind turbines are often used in hybrid systems (e.g. with diesel and solar power).

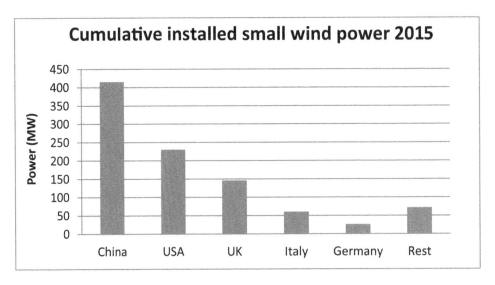

Figure 2.26 The five countries with the most power installed in small wind turbines. China has 44% of the installations.

Source: Small Wind World Report 2017, WWEA [14]

Figure 2.27 A Vertical-Axis Wind Turbine (VAWT), 12 kW, designed and constructed at Uppsala University, Sweden [15]

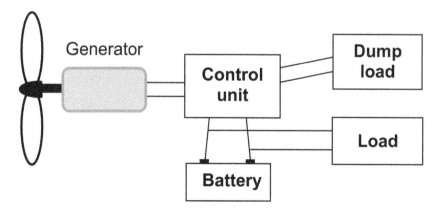

Figure 2.28 A typical small wind power stand-alone system with direct current (DC).

In China, off-grid installations represented 97% of the market in 2009. More than two million people there still lack access to electricity grids. In the United States, Alaska is not covered by the conventional grid and there is a popular movement also elsewhere in which several thousand people want to live "off-the-grid" (OTG). The term can refer to living in a self-sufficient manner without reliance on one or more public utilities, such as the electrical grid.

Electrification of remote locations can often be done comparatively cheaply and easily with small wind systems, if wind conditions are suitable. If an all-year supply of electricity is needed where the distance to the existing grid is long and wind conditions are good during winter when the solar resource is poor, then a local supply by wind energy is often the best alternative.

In many small wind systems, the generator primarily charges the batteries. Therefore, the voltage output published by the small wind turbine manufacturer is often the nominal DC voltage of the batteries, even if the generator will actually generate AC, which is rectified. The control unit protects the batteries against overcharge (and preferably also against too-deep discharge). Excess energy is usually diverted to a dump load, when the batteries are fully charged. It is important not to store the battery in a place that can become hot. High temperatures shorten battery life.

In small systems it is often better to use DC loads, if possible. An inverter can convert DC to AC, but it is expensive and gives losses. However, sometimes it is necessary with AC for some loads. Regardless whether DC or AC is used, correct fusing is always important.

A small wind turbine can also be grid connected; see Figure 2.29. This is usually done to lower the utility bill. The electrical power is in this example converted to the grid voltage and frequency by an inverter, but some small wind turbines have an AC generator that can be connected without any inverter. However, the cost per produced kWh is higher for a small wind turbine than for larger machines.

Normally it is recommended to use only the types of towers and controllers specified by the wind turbine manufacturer. If you use a tower of your own design, unexpected dynamic problems (vibrations) may occur.

Be aware that the quality and robustness of small wind turbines vary significantly between different models and manufacturers. Different sites will also impose different demands. Consequently, some users have very good experiences using small wind turbines, while others have had bad experiences.

Note also that small wind systems often have relatively high installation costs (unless e.g. a homeowner installs the machine himself, and does not count the cost for this).

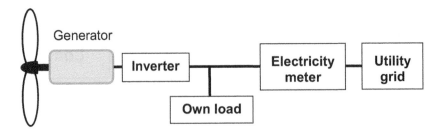

Figure 2.29 Small, grid-connected wind power system.

While mechanical wind pumps still exist, other newly developed wind pumping technology uses electricity [16] [17] [18]. It gives possibility to optimize the siting, and the wind turbine can be located separated from the water supply. Wind pumping systems are, for example, used for village water supply, and similar technology could also be used for desalination, where a pump presses water through a membrane (reverse osmosis) [19]. An example that has been tested by one of the authors is to use a wind turbine with a permanent magnet generator that reaches approximately 400 V AC and 50 Hz at maximum speed (limited for example by passive pitch control) in combination with an asynchronous (induction) motor rated 400 V AC and 50 Hz to drive a pump. At lower wind speeds, the voltage, frequency and water output will be less. If the generator has low inner inductance, no capacitors are needed. However, if the pump motor has a motor protection, that may disable operation at low voltage and frequency.

2.2.10. External conditions

To the buyer of a wind turbine, it is important to ensure that the machine is designed to cope with the external conditions that can be expected at the current site.

For some external conditions, such as ambient temperature, it is easy to determine the extreme values of the site and compare them with the manufacturer's specification. If the temperature on the site can be higher than the specification, a typical problem is that the cooling system of the machine may be insufficient, but the manufacturer can sometimes solve this with a retrofit (e.g. a larger cooler or automatic reduction of the maximum power produced during the few hours with extremely high temperature). However, if temperatures below the specification can be expected on the site, a retrofit solution is often not enough. For example, at low temperatures it is important to use the right kind of steel in all load-carrying components. Otherwise the impact strength of the steel can be insufficient.

For some other external conditions, which are not always easy to learn for the site, a more extensive description is given next.

2.2.10.1. Extreme wind speed

The survival wind speed (the wind speed that the wind turbine can withstand without severe damage) of the turbine is normally compared with the expected 50-year extreme wind speed at hub height on the site, and shall not be less than that value. When comparing such values, it is important to consider the height and averaging time. Meteorological data normally refer to a height of 10 meters above the ground, while wind turbine data normally refer to hub height.

It is possible to calculate the 50-year extreme wind speed from long-term wind measurements, for example with the software Windographer.

If the survival wind speed of the wind turbine is too low, there is a high risk that the machine will be destroyed during extreme wind speeds. This is especially important to consider for areas prone to tropical cyclones.

In Sweden, where there are no tropical cyclones, the relation between survival wind speed and average wind speed specified in the standard classes of the IEC safety standards is normally sufficient. This means that here the selection of the appropriate standard class can typically be based on the average wind speed of the site (which is usually known) without further investigation of the survival wind speed.

Although a wind turbine may have been designed for a certain extreme wind speed, there is no 100% guarantee that it will in practice withstand events with such high winds. For example, during hurricane-force wind speeds there is often flying debris that can damage small wind turbines, no matter what wind speed they were designed for.

2.2.10.2. Turbulence

Turbulence is important, because it can cause fatigue damage on a wind turbine. The lower the turbulence level, the smaller the turbulence-induced fatigue loads. Another issue associated with high turbulence can be rapid wind direction changes, which can be a problem for some turbines.

As an example, if a small wind turbine is installed on a too-short tower near buildings or trees, it is subject to high turbulence levels when it is in the zone of disturbed flow. Therefore, such bad installations shall be avoided to ensure a long life of the wind turbine, if possible.

For applications in which high turbulence cannot be avoided, it is important to select a machine that can cope with those conditions.

2.2.11. Installation

Large and medium wind turbines are normally planned carefully and installed by professional crews using mobile cranes. The following text is therefore focused on small wind turbines, which are sometimes installed by the purchaser.

A wind turbine must have a clear view in the direction of the prevailing winds to perform efficiently. Turbulence, which reduces performance and works out the turbines harder than a smooth air stream, is highest close to the ground and decreases with height. Wind speed also increases with height above the ground. As a general rule of thumb, you should install a wind turbine on a tower such that the rotor disc is at least 10 meters above any obstacles within 100 meters. Smaller turbines typically go on shorter towers than larger turbines. A 250-watt turbine is often, for example, installed on a 10- to 15-meter tower, while a 10-kW turbine will usually need a tower of 24–36 meters. It is not recommended mounting wind turbines on buildings in which people live, because of the risk for problems with vibration and noise. The wind pattern is also complicated around a building, and production can become low.

Towers and particularly guyed masts can be hinged at their base and suitably equipped to allow them to be tilted up or down, for example using a winch. This allows all work to be done at ground level. Some towers and turbines can easily be erected by the purchaser, while others are best left to trained professionals. Aluminium towers should be avoided because they are prone to developing cracks. Towers are usually offered by wind turbine manufacturers, and purchasing one from them is a good way to ensure proper compatibility.

2.2.12. Environmental issues

Operation of wind turbines does not cause air or water emissions. Wind is pollution-free electricity and can help us to reduce the environmental damage caused by other forms of power generation. However, wind systems, like all other energy technologies, have some environmental impacts.

There are a few environmental risks associated with waste treatment of wind systems. Small wind systems often contain large batteries, which must be properly recycled after use.

Used oils are classified as hazardous waste and have to be taken care of in a specific manner. Wind turbine blades can contain PVC, which can pollute the air if combusted.

Birds and bats occasionally collide with wind turbines, as they do with other objects, but there are ways to reduce such risks (e.g. by automatically shutting down wind turbine operation during those short periods when there is a high risk of bat collisions). Wind turbines give visual impacts, which can be minimized through careful design of the installation. Rotating blades causing shadow flicker can be an issue, mainly at higher latitudes [21]. A shadow flicker calculation is usually also part of the permitting process for large turbines.

Noise was a problem with some early wind turbines, but it has been largely eliminated. Aerodynamic noise has been reduced by improved blade design. A small amount of noise is generated by the mechanical components of the turbine, but very little of this can normally be heard from the ground.

Life cycle assessments of wind turbines have shown that it normally takes only a few months of operation for a wind turbine to "pay back" the energy consumed for manufacture, etc,. of it.

2.2.13. Safety concerns

A properly designed, installed and maintained wind system poses very little risk to the public. Thousands of wind turbines are installed in the world, and their safety track record is for the most part good. Trees are much more likely to fall or drop branches than an ordinary wind turbine, but no minimum security distance is required for trees. However, there are safety concerns associated, for example, with transport, installation, maintenance and demolition of wind turbines, just like with other tall structures.

Serious accidents have occurred, for example when people have fallen while working at high heights. Always use proper personal safety devices when working at a high level. Be careful not to drop anything, and never position yourself where there is a risk of falling objects. Also, don't leave a wind turbine in a potentially dangerous state, for example during maintenance.

For personnel working at high heights, the training normally includes evacuation (e.g. of a wounded colleague) from a high height.

A tilt-down tower that allows maintenance to be done in a lowered position is one possibility to avoid working at high heights. However, this introduces other risks when raising or lowering the tower.

In the European Economic Area, special requirements exist regarding a control device whereby the machinery can be brought safely to a complete stop and also regarding documentation. It is actually unclear if some of the small wind turbines on the market fulfil these requirements, even if the manufacturer has applied the CE Mark.

2.2.14. Consumer labels and certification

To help buyers of small wind turbines make an informed decision, standards have been developed for consumer labels. The idea is that the label shows results based on standardized testing in a condensed and comparable form, regardless of where the testing has been conducted. Presently there are two systems for small wind turbine consumer labels: the IEC consumer label (Fig. 2.30), which is based on testing only and done according to international standards, and the AWEA consumer label, used mainly in North America, for turbines that are certified to the AWEA small wind turbine standard.

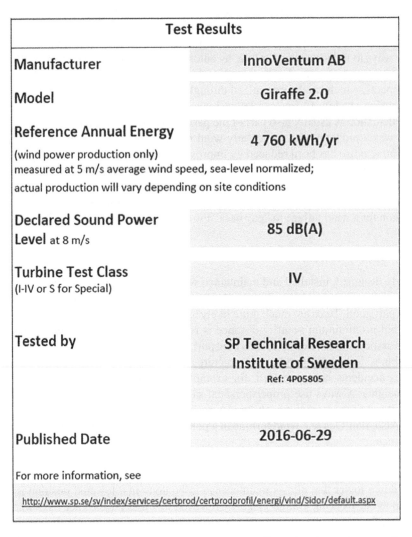

Test Results	
Manufacturer	**InnoVentum AB**
Model	**Giraffe 2.0**
Reference Annual Energy (wind power production only) measured at 5 m/s average wind speed, sea-level normalized; actual production will vary depending on site conditions	**4 760 kWh/yr**
Declared Sound Power Level at 8 m/s	**85 dB(A)**
Turbine Test Class (I–IV or S for Special)	**IV**
Tested by	**SP Technical Research Institute of Sweden** **Ref: 4P05805**
Published Date	**2016-06-29**
For more information, see http://www.sp.se/sv/index/services/certprod/certprodprofil/energi/vind/Sidor/default.aspx	

Figure 2.30 Example of an IEC consumer label for the small wind turbine shown in Figure 1.3, with test results for it included in Table 2.2 [20]. At the test site in Malmö, the solar panels will produce an additional 5900 kWh/year (approximately).

Test results displayed on IEC and AWEA labels are comparable, except for sound, which is explained on small-wind.org/labels/.

It is standard practice for large wind turbines to have a type certificate, which means that an accredited third-party certification organization has reviewed the compliance with standards (normally in the IEC 61400 series). Also some small wind turbines can be found that have a type certificate to the international small wind turbine standard IEC 61400-2 or to other schemes like MCS in the UK. In Japan, ClassNK certification is used for small wind turbines. Regarding Danish type approval of small wind turbines, see the website links listed later.

Table 2.2 Small wind turbine test results for consumer labels (some numbers have been rounded)

Manufacturer	Model	Swept area [m²]	Power at 11 m/s [kW]	Reference annual energy at 5 m/s [kWh]	AWEA-rated sound pressure level [dB(A)]	IEC declared sound power level [dB(A)]	Turbine test class [I-IV]	Max C_p	Cut-in wind speed [m/s]
Pika Energy	T701	7.1	1.5	2420	38.3		II	0.31 (at 8.5 m/s)	3.0
Xzeres Wind Corporation	Skystream 3.7	10.9	2.1	3420	41.2		II	0.29 (at 7–8 m/s)	3.0
Eveready Diversified Products (Pty) Ltd.	Kestrel e400nb	12.6	2.5	3930	55.6		II	0.30 (at 7.5–9 m/s)	4.0
InnoVentum AB	Giraffe 2.0/Windspot 3.5	12.9	2.7	4760		85	IV	0.32 (at 5.5–9 m/s)	2.5
Sonkyo Energy	Windspot 3.5	13.2	3.2	4820	39.1	84.6	IV	0.32 (at 6–7 m/s)	2.5
SD Wind Energy (formerly Kingspan/Proven)	SD6 (KW6)	24	5.2	8950	43.1		II	0.36 (at 8 m/s)	2.5
Bergey Windpower Co.	Excel 6	30	5.5	9920	47.2		II	0.31 (at 4–4.5 m/s)	2.5
Bergey Windpower Co.	Excel 10	38	8.9	13,800	42.9		II	0.30 (at 6–9 m/s)	2.5
Xzeres Wind Corporation	442SR	41	10.4	16,700	48.5		II	0.37 (at 7.5–8 m/s)	2.5
Lely Aircon B.V.	LA10	45	9.6	17,500	41.1		II	0.38 (at 6.5–7.5 m/s)	3.5
Britwind	H15 Class II (20 kW version)	85	14.5	31,400		89.9	II	0.39 (at 6–6.5 m/s)	3
Britwind	H15 Class IV (14 kW version)	135	11.8	38,000		90.7	IV	0.39 (at 5 m/s)	3
Dakota Turbines	DT-25	129	23.9	47,800	33.8		II	0.33 (5–8 m/s)	2.5
Lely Aircon B.V.	LA30	135	27.2	48,800	49.8		II	0.35 (at 7 m/s)	3.5
Solid Wind Power A/S	SWP25–14GT20	154	25.2	57,000			III	0.41 (at 6 m/s)	3.0

Figure 2.31 Free-falling water in a small stream. The energy in the stream can be captured in a hydropower plant.

2.3. Hydropower

In hydropower the energy of flowing or falling water is utilized (Fig. 2.31). It is solar energy that was captured when water by evaporation from land and water surfaces is condensed in the clouds and dropped off in higher areas. The water passes a turbine which is given rotation. The turbine drives a generator that converts kinetic energy into electrical energy.

Hydropower can be used directly as a mechanical force. This was common in the past, for example in mills that ground grain between rotating millstones. But hydropower became more important when we learned to generate electricity and transmit energy with power lines. The definition of small hydro varies between 1 and 50 MW but typically it is hydropower projects with power up to 10 MW. About 12% of hydropower plants are considered to be small.

Hydroelectricity is one of the cheapest methods to produce new electricity. The hydro station consumes no water or fuel, so the costs of operation are low. When a hydroelectric station is constructed, the project typically produces no waste, and has a considerably low emission of greenhouse gases. The economic lifetime of a hydroelectric station is long. Some stations have been in operation for 50–100 years.

2.3.1. Hydropower history

Some of the earliest innovations in using water as an energy source were conceived in China during the Han dynasty between 202 BC and AD 9. Trip hammers powered by the vertical-set waterwheel were used for example in early paper-making.

The oldest water mills known in Sweden were in operation in the AD 300s, more than 1600 years ago. Water was used for many purposes. It could, for example, be used for grinding grain or sawing and planing wood. Figure 2.32 is of Brunnsbacka sawmill which dates

Figure 2.32 Brunnsbacka sawmill. It was used for sawing and planing wood in the 1700s and 1800s. The mill is located by a creek in the Swedish province of Småland.

Source: G. Sidén.

from the 1700s. Hydropower became an important source of energy for the early industry. Many of our mills have been built by rivers, where they used the water for driving processing machines.

Some of the most important developments in hydropower technology happened in the first half of the nineteenth century:

- In 1849, the British–American engineer James Francis developed the first modern water turbine – the Francis turbine – still the most-used water turbine in the world today.
- In 1880, an American inventor, Lester Allan Pelton, got a patent on the Pelton wheel, an impulse water turbine.
- In 1913, the Austrian professor Viktor Kaplan developed the Kaplan turbine – a propeller-type turbine with adjustable blades.

The world's first hydroelectric project was to give power to a single lamp in the Cragside country house in Northumberland, England, in 1878.

Four years later, the first plant to serve a system of private and commercial customers was developed in the state of Wisconsin, USA. Within 10 years, hundreds of hydropower plants were in operation.

The same year, 1882, industrial history was written in Rydal at the river Viskan in Sweden. For the first time a hall in a textile industry were lightened with lamps powered from a hydroelectric generator. It solved a big problem; kerosene lamps that lit up production halls earlier carried significant fire hazards. The generator power was 3 horsepower (approximately 2.2 kW) and gave electrical energy to two arc lamps, which lit up the factory.

Figure 2.33 The company Vattenfall operated Rydal's hydropower plant; it has now been preserved as a historic building.

Source: B.-E. Sidén.

A new power station in Rydal began operating in 1916 and was rebuilt in 1927 after large floods. The new power plant (Fig. 2.33), with the power to generate 680 kW, was driven by a Kaplan turbine.

In 1895, the world's largest hydroelectric development of the time, the Edward Dean Adams Power Plant, was created at Niagara Falls.

2.3.2. Head and flow gives power

From a basic understanding of available potential energy that can be converted to kinetic energy, we can understand that large water flows and heads (differences in the levels of rivers) are the factors that are important for utilization of hydropower. For best utilization, the height differences must be concentrated to the power plants. Few hydropower plants are built where there is a large natural falls. Many rivers have a relatively flat path and the drop for a fairly long stretch needs to be concentrated to the plant. This is done by damming up the river or transporting the water via tunnels, channels or tubes to the plant. Examples are shown in Figure 2.34. The smallest impact on nature occurs if the river is relatively narrow and surrounded by steep slopes.

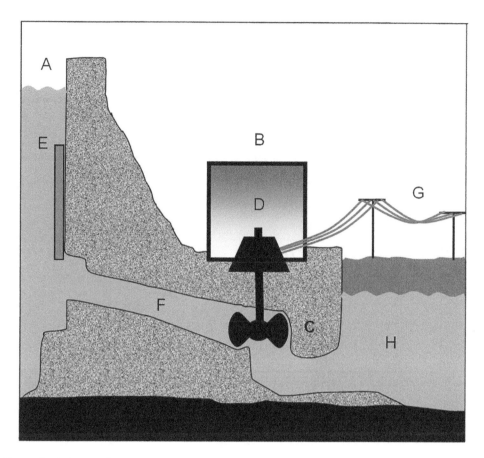

Figure 2.34 Parts of a hydroelectric plant: A = reservoir, B = powerhouse, C = turbine,
    D = generator, E = intake with grid and gate, F = penstock, G = power line,
    H = receiving river or lake.

The water goes through the intake gate, intake cover and intake tunnel to a turbine. The intake grid prevents larger objects such as ice, timber and fish from entering the turbine. With the intake gate, the water flow is controlled, so that production can be balanced against the demand for electricity. Most common turbine types are Pelton, Francis and Kaplan. These three main types are suitable for different vertical heights, water flows and power, see Figure 2.35. Both the Pelton and Kaplan turbines have high efficiency, even at part load.

In the Kaplan turbine the blades are adjusted for different water flows. The turbine shaft drives the generator which then converts the turbine's kinetic energy into electricity. The voltage the generator leaves is often raised by a transformer to match the voltage in the grid. The efficiency of hydropower turbines is high, often around 90%.

There are also other forms of hydropower, such as Archimedes' screws (which are also useful as pumps) and Ampair UW100 (Fig. 1.2).

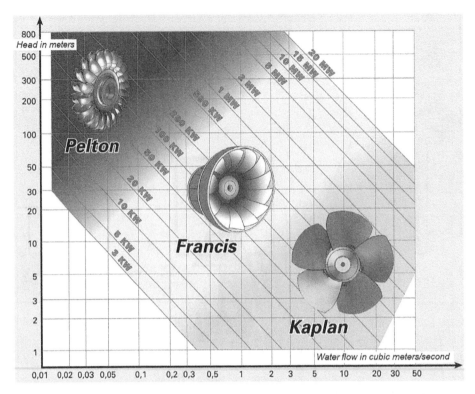

Figure 2.35 Pelton, Francis and Kaplan are the most common types of turbines in hydroelectric plants. They are suitable for different vertical drops, water flows and
power.

Source: Gugler GmbH.

2.3.3. Control of hydropower

Hydropower is very important for the control of power in the electric grid. In Sweden, peak
power demand normally occurs in the early-morning hours on chilly days in wintertime. The
lowest consumption is during weekend nights in July. Then the consumption is only one third
of the largest. In average the power needs are less than half of installed power. Normally
around 60% must be downregulated or switched off.

Electrical energy that is produced to the grid must be used at the same time. Otherwise an
imbalance in the electric grid occurs. The regulation of the power system is enormously important, and hydropower is the power source that is easiest to control. That occurs through controlling water intake by opening or closing the gates to achieve the desired power. The setting
can be done quickly, which is needed sometimes. In some minutes, full production is reached.

Normally a minimum water flow should be maintained in the river, but water quantities over this can be stored in the reservoir and used in exactly the amount needed for the
moment. Also for other energy sources such as wind, hydropower is a valuable complement.
When the wind blows, much water flow can be downregulated, and the energy is stored in
reservoirs for later use.

2.3.4. Dams and water reservoirs

Electricity is difficult to store on a large scale. However, the water that gives hydropower can be stored in reservoirs along the rivers. In Sweden, the reservoirs fill up when the snow melts and during autumn rains. But the need for electricity is highest during cold winter days with a lot of heat pumps in operation and hot summer days when a lot of air conditioning is used. Then the stored water can be used for electricity generation.

The hydroelectric reservoirs can store water for a short time, for example from night to day or some days, to handle varying inflow and electricity use, as well as level out the production from other sources.

It is also possible to use hydro for seasonal storage, which stores the energy for several months. In Sweden the storing capacity in the hydroelectric system is about 34 TWh. That is about one fourth of the total yearly electricity generation in Sweden.

In Norway the storing capacity is much higher. They have more than 1000 reservoirs in their hydropower system. They can store 87 TWh [22], or about 70% of their total yearly electricity generation.

In the future Norway can function as Europe's blue-green battery, using the flexibility of the Norwegian hydropower system for large-scale balancing and energy storage [23]. Today Norway has HVDC (high voltage direct current) sea cables to Denmark and the Netherlands but soon cables will be installed to Germany and the UK. Then the Norwegian storage capacity can be used for much of Northern Europe.

Energy can also be stored in a pumped hydropower system. This can occur for both small and large systems. For a pumped hydro storage, two reservoirs of different heights are needed. Energy storage technology is covered in Chapter 3, and an example of pumped hydro is described in the final section of Chapter 6, which discusses the island of El Hierro (including a sample calculation of hydropower).

2.3.5. Hydropower in the world

Hydropower is the largest renewable source for electricity generation in the world. It provided 4185 TWh, or 16.4%, of the global generation in 2017. Table 2.3 lists the installed capacity in different countries. The hydropower installed globally rose to 1267 gigawatts (GW) at the end of 2017, including 153 GW of pumped storage. During the year, installations increased by 1.6%. In total, 21.9 GW of new capacity was added, including 3.2 GW of pumped storage.

The use of hydropower expanded greatly in North America and Europe until around 1980. Now this has changed. In 2017 the most expansive hydroelectric development occurred in East Asia and the Pacific, with 9.8 GW of capacity added, followed by South America (4.1 GW), South and Central Asia (3.3 GW), Europe (2.3 GW), Africa (1.9 GW) and North and Central America (0.5 GW) [24].

After construction of the hydropower project, the Three Gorges dam, China is now the largest hydroelectric power country in the world. The dam goes over the Yangtze Kiang, the Yellow River, and is 2.3 kilometers long, 185 meters high and 15 meters wide, and contains 16 million cubic meters of concrete. It is the largest hydroelectric dam in the world, and the water reservoir reaches 400 kilometers up the river. In total, 1.3 million people had to move from the dam area.

Table 2.3 Countries in the world with the largest hydropower installation [24]

Country	Installed capacity 2017 (GW)	Share of world installation (%)
China	341	26.9
USA	103	8.1
Brazil	100	7.9
Canada	81	6.4
Japan	50	3.9
India	49	3.9
Russia	48	3.8
Norway	32	2.5
Turkey	27	2.1
France	26	2.1
Italy	22	1.7
Spain	20	1.6
Switzerland	17	1.3
Vietnam	17	1.3
Sweden	16	1.3
Venezuela	15	1.2
Austria	14	1.1
Mexico	12	0.9
Iran	12	0.9
Colombia	12	0.9
Other	252	19.9
World	**1267**	

Table 2.4 Different categories of small hydropower plants

Hydro category	Power range	Annual generation (capacity factor 0.5)	Number of powered households (4000 kWh each)
Small	1 MW – 10 MW	4.4–44 GWh	1100–11,000
Mini	100 kW – 1 MW	0.44–4.4 GWh	110–1100
Micro	5 kW – 100 kW	22–440 MWh	5–110
Pico	0 kW – 5 kW	0–22,000 kWh	0–5

2.3.6. Small hydropower

The limit for what is called small hydropower plants is usually set at the power of 10 MW. However, there is no generally accepted limit. In Sweden, the limit is considered to be 1.5 MW and in the USA, Canada and China it can be up to 50 MW. A further categorization is shown in Table 2.4.

Three different designs are used for small hydropower plants:

*   Run-of-the-river: The streaming water is used to generate power without changing the water flow in the river.
*   Storage: These power plants have a possibility for water storage, through an existing or newly built dam. The generation can be adapted to the power needs.

- Pumped storage: There it is a possibility to pump the water to an upper reservoir and store it until needs occur. This type of design is not common for small hydropower plants.

Mini-, micro- and pico-power plants do not normally have dams and therefore they are run-of-the-river plants. After use, the water returns to its natural flow. This minimizes impacts to the surrounding environment and nearby communities.

2.3.7. Global overview of small hydropower

The globally installed small hydropower capacity is estimated to have been 78 GW in 2016. That is an increase of approximately 4% compared with data from 2013. The total estimated small hydropower potential has also increased since a report from 2013 to 217 GW, an increase of more than 24%. Thus, approximately 36% of the total global potential has been utilized 2016 (see Fig. 2.36). The size definition for small hydropower plants can be misunderstood, for a "small" hydro system could actually produce electricity for 11,000 average European homes, which can be considered significant.

Small hydropower gives approximately 1.9% of the world's total electricity production. Approximately 11.5% of total hydropower (including pumped storage) is considered small (<10 MW).

China dominates the small hydro sector. About 29% of the world's total potential exists in China. The country also has the largest share of the installations so far, 51% of the country's potential.

The four largest countries thereafter – Italy, Japan, Norway and the United States – account for 16% of the world's total installed capacity.

In East Asia, besides China, Japan also has great potential for expansion. So far, about one third of the potential has been exploited in Japan. In Central Asia, only about 4% has been used, but opportunities are great.

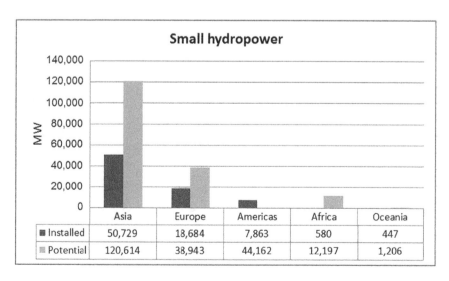

	Asia	Europe	Americas	Africa	Oceania
■ Installed	50,729	18,684	7,863	580	447
▦ Potential	120,614	38,943	44,162	12,197	1,206

Figure 2.36 Installations and estimated potential of small hydropower (<10 MW) in different continents at the end of 2016 [25].

Table 2.5 Countries with a potential more than 200 MW for small hydropower and their installed power [25]

Continent	Country	Potential (MW)	Installed (MW)
Asia	China	63,500	39,800
	India	11,914	2119
	Japan	10,270	3545
	Vietnam	7200	1836
	Turkey	6500	1156
	Kazakhstan	2707	78
	Pakistan	2265	287
	Laos	2000	12
	Philippines	1975	101
	Nepal	1430	131
	North Korea	1500	159
	Turkmenistan	1300	5
	Afghanistan	1200	80
	Uzbekistan	1180	71
	Kyrgyzstan	900	42
	Sri Lanka	873	308
	Indonesia	770	229
	Thailand	700	108
	Malaysia	500	18
	Armenia	396	282
	Azerbaijan	392	13
	Cambodia	300	1
Africa	Kenya	3000	32
	Ethiopia	1500	6
	Ghana	1245	0
	Mozambique	1000	2
	Angola	861	13
	Nigeria	735	45
	Cameroon	615	1
	Tanzania	400	25
	Sierra Leone	330	11
	South Africa	247	50
	Uganda	200	35
Europe	Russia	-	214
	Norway	7676	2242
	Italy	7073	3173
	France	2615	2021
	Spain	2185	2104
	Greece	2000	223
	Albania	1963	65
	Germany	1830	1826
	Austria	1780	1368
	Sweden	1280	1280
	UK	1179	274
	Ukraine	1140	82
	Bosnia/Herzeg.	1000	36
	Switzerland	859	859
	Portugal	750	372
	Poland	735	288
	Romania	730	598
	Finland	590	314
	Bulgaria	581	291
	Slovenia	475	157
	Czech Republic	465	334
	Serbia	409	46
	Moldova	300	0
	Macedonia	260	60
	Belarus	250	16
	Slovakia	241	82
North/Central America	USA	6366	3676
	Canada	1113	1113
	Mexico	470	470
	Honduras	385	75
South America	Brazil	-	1023
	Colombia	25,000	250
	Chile	7000	175
	Peru	1600	391
	Argentina	430	78
	Ecuador	296	94
	Uruguay	232	0

In South Asia, the largest potential assets are in India, of which about 28% has been utilized. There are good opportunities also in Pakistan, Nepal and Bangladesh.

In Southeast Asia there is good potential in Vietnam (25% utilized), the Philippines, Laos and Indonesia.

Asia has had the largest increase in installed capacity, and it has increased by 33% in 2013–2016. In installed capacity, the increase was 4462 MW.

The proportion of small-scale hydropower that is utilized in Africa is about 5%. In East Africa, Kenya, Ethiopia and Mozambique have the greatest potential, but the proportion that has been expanded is very low. Angola and Cameroon have the greatest potential in Central Africa and in West Africa it is Ghana and Nigeria.

The largest installed power in Africa is Congo-Kinshasa and South Africa, with approximately 50 MW each. Africa had the second largest increase in 2013–2016. It was 10%, or 54 MW.

In Europe about 50% of small hydropower potential has been utilized. In some countries almost the entire potential has been utilized, for example in Sweden, Germany and Switzerland. Russia has the biggest potential (data lacking, however), as well as Norway and Italy.

In North America, the United States and Canada have great potential for small hydropower. In South America, the potential is largest in Brazil (data lacking, however) followed by Columbia and Chile.

2.3.8. Mini power plant in Sweden

At the small stream Lillån, a tributary to the river Ätran, there have been mills since the 1600s. The power from the water flow has also been used for other purposes. Among other things, there were previously two sawmills on the received power from the water. In 1923, a mill was converted to an electrical power plant and the small industrial buildings got electricity. The old power plant can still be used. It is equipped with a turbine, synchronous generator and regulator for island operation. A few years ago, it was used when the public grid was out of operation a week after a storm.

The old power plant did not use the full potential of the site. The owner decided to build a new one, which has higher power and is more efficient (Fig. 2.37). The old Francis turbine had a high efficiency at the best water flow, but the new Kaplan turbine can adjust the turbine blades so it gets a high efficiency at different flows, and the resulting average efficiency becomes much higher (Fig. 2.38).

The old dam could be used but a new penstock had to be installed (Fig. 2.39). The dam was also refurbished with an automatic floodgate. The function of the dam is only to collect the water flow so that it goes to the penstock. It has no storage function, so the power output must always be adapted to the natural flow. The average flow during the year is 1.6 m³/s, but it can also be twice as large, and then the old power plant can be run in parallel.

The cost of the new power plant was approximately € 1.2 million. A total of 40% of the cost went to machine equipment and 60% to building and construction work. It is expected that the repayment period with the current electricity price could be 20 years. But it is a solid building with good machine equipment so it is expected that the lifetime, even for the penstock, will be up to 100 years, making it a profitable project.

The steep stream has always been a stop for migrating salmon. But the eel has walked up the stream. To help the eels pass the plant on their way upstream, a collector for small eels

Figure 2.37 The new hydropower station built in 2016 at the small stream Lillån, a tributary to the river Ätran in Sweden, will double electricity generation compared with the old station.

Source: G. Sidén.

Some data for the mini plant
Kaplan turbine
Head: 22 meters
Max. water flow: 3 m3/s
Penstock: Diameter 1.5 meters
Length: 98 meters
Nominal power: 600 kW
Production/year: 2000 MWh

Figure 2.38 The generator: a key component of the new power plant located on a tributary of the river Ätran in Sweden
Source: G. Sidén.

has now been installed at the outlet. Every day in the summer, the caught eels are carried and released above the pond. In 2017, more than 1800 eels were transported up. There is also a bypass where the eels and also other fish can migrate down without being damaged. The new work has become a good improvement for the river fauna.

Figure 2.39 The new penstock is 98 meters long and made from Fiberglass-reinforced plastic. It is completely buried in the ground and thus protected from external influences.

Source: Jan-Åke Jacobson.

In the environmental trials, part of the old riverbed was required to have a water mirror. It has also been remedied, so that the visual impression for the surroundings has also improved with the new power plant.

2.3.9. Pico hydropower

Even a very small water flow in isolated places can be used for electricity generation. Figure 2.40 illustrates how a simple plant can be constructed. The intake must be below the water surface. One can therefore arrange to have a small dam or water tank. Properly designed, this reduces the risk that debris, such as twigs and leaves, will enter the penstock. The intake should also have a relatively tight grille to prevent fish and small animals from entering.

The annual production corresponds to power in the range of 20 W–3.5 kW. As with all hydropower, the available water flow varies greatly during the year. But production is still more predictable than it is from solar and wind power. Production can also be as great at evening and nighttime as during daytime.

Figure 2.41 shows a turbine where the turbine flow can be controlled by a pipe parallel to the turbine.

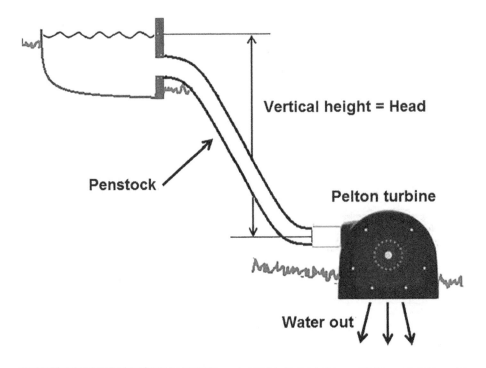

Figure 2.40 A simple layout for a micro or pico hydropower plant [26].

Figure 2.41 A pico-sized Pelton turbine can handle the supply to a regular European home. This turbine has two jets, which are fed by the pipe network.

Source: Courtesy of Hugh Piggott and Michael Lawley, Powerspout.com [26].

Table 2.6 Estimated production for three different turbine types for different flows and heads [26]

Estimated annual production (kWh)

	1 l/s	2 l/s	4 l/s	8 l/s	16 l/s	32 l/s	64 l/s
2 m				450		2500	
4 m		200	500	1200	2300	5000	8700
8 m	200	500	1200	2400	5300		
16 m	500	1200	2400	4500	10000		
32 m	1200	2400	5000	8500	15000		
64 m	2400	5000					
128 m	5000	10600					

| PELTON TURBINE | TURGO TURBINE | LOW HEAD TURBINE |

A typical annual consumption in a home in Europe is about 4000 kWh, and that is possible to produce with many of the flow/head options in Table 2.6. Pico hydropower can be used both for on-grid and off-grid applications. On-grid systems will normally need a "grid-controlled inverter" as used in a grid-connected solar PV system. Off-grid systems may need a battery to store the energy, and use an inverter for AC power. It can also be combined with other generation, such as a fuel-driven backup generator.

2.3.10. Environmental impact

Hydropower is a clean and renewable energy that does not release any major emissions into the air or water. When it replaces fossil fuels that contribute to the greenhouse effect, climate change is reduced. Hydroelectric power can cause limited greenhouse gas contributions when areas with plants that absorb carbon dioxide are reduced. And the decomposition of organic matter in reservoirs emits carbon dioxide. If the nutritional supplements are sufficiently large and the oxygen supply is small, emissions of methane gas can occur, which produces a much larger greenhouse effect. It is mainly from power plant dams in tropical areas that emissions of methane gas are reported.

The reservoir and the operation of the dam can also result in residential areas having to be abandoned and people forced to move. Farmland and forest areas can also be lost. The physical effects of a pond and reservoir, the operation of the pond and the use of the water can change the environment over a much wider area than the area covered by the facilities themselves.

In the construction of power plants and dams, considerable resources are used, mainly steel and concrete. The production of these materials leads to emissions of carbon dioxide,

Figure 2.42 Salmon ladder at a hydropower plant.

Source: G. Sidén.

sulfur and nitrogen oxides. When soil and rock masses are moved at the site, emissions are generated from excavators and trucks.

2.3.11. Landscape and nature

Large areas above the hydropower dam can be flooded, while the river downstream is completely or partially emptied. New shorelines are formed and areas that may be under water in autumn and winter are dry in the summer and large, empty areas are formed.

The construction of reservoirs, dams and power plants is an extensive process. Transformer stations, power lines and roads must also be built. Everything leads to a changed landscape and has a great impact on nature, even if one tries to incorporate residues from excavation and blasting into the natural topography.

Water flow to and from the plant is often accomplished in long tunnels, so the flow in a part of the river is reduced or completely disappears. With construction of a power plant, the opportunities for people to experience an environment with flowing water is reduced.

The river is also affected downstream from dams and power plants. Living conditions for the species that live by flowing water will change. Plants that are dependent on annual floods disappear or decrease, for example, if the spring flood disappears.

2.3.12. Affected animal species

The changes caused by hydropower have led to poor conditions for some species, but other species may also benefit. The significant changes to water levels in reservoir areas affect plants and animals in the surrounding environment. River shores lose much of their biodiversity.

Most severely affected are fish, for example salmon, trout and eel. Because of dam structures, the fish can no longer swim back to their original spawning areas. The different water levels also affect the availability of fish feed and plants in the beach zone.

Approaches to solving these problems are, for example, the construction of fish ladders (Fig. 2.42) and other ways to help fish get around or over the plants to spawning areas upstream and bypasses for downstream migration. The problems can also be mitigated by active fishing management, for example the release of salmon fry.

2.4. Geothermal power

Geothermal energy is different from other renewable energy sources because most of the energy comes from the earth and not from the sun. The heat leaking through the crust is partly energy stored from when the earth was formed, but also energy that is newly formed by nuclear reactions in the earth's interior. The energy is stored in the rock and the water that fills the pores and fractures. A limited portion of the geothermal energy reaches the surface in the form of hot springs or geysers. The energy can be extracted as hot or evaporated water from deep boreholes and used for power generation (Fig. 2.43).

Even so-called hot dry rock mining technology (HDR) is used. Cold water is pumped down and heated up in the deep rock layer, after which the hot water is returned to the surface. Geothermal heat can be used directly for heating or, if the temperature is high enough, converted to electrical energy. Sometimes the temperature is raised further with heat pumps.

Figure 2.43 The Nesjavellir Geothermal Power Station (NGPS) is the second-largest geothermal power station in Iceland. The electric power is 120 MW, and the yearly production 1 TWh. It also delivers around 1100 liters of hot water (82–85 °C) per second, serving the space heating and hot water needs of the capital region.

The assets of geothermal energy are considerable, but only a fraction can be extracted depending on the depth of the usable temperature, geological composition of the bedrock and internal processes in the crust. Areas with active volcanoes or a thin crust can be easier to exploit energy. The prospects for extraction are best at the borders between the continental plates. A continental plate is a coherent part of the earth's crust. In Sweden, for example, which is in the middle of such a plate, prospects are bad for geothermal energy.

The geothermal gradient describes how the temperature increases with depth. The temperature increases in average 3° C per 100 meters. If the surface temperature is 15° C, the average temperature at 2000 meters depth is 65–75° C. But because of asymmetries in the crust and underlying layer, the gradient varies sharply. In some areas, as in the Swedish bedrock, it is only 1.5° C per 100 meters, while in Iceland it can reach 100° C per 100 meters.

Geothermal energy is the fifth largest source of renewable energy after bioenergy, hydropower, wind power and solar power. Geothermal energy for electricity generation has grown substantially (Fig. 2.44). For 2015 and 2017, the average increase was 4.9% per year. In 2017, the total installations in 24 countries were 12,913 MW, and the annual production of electrical energy is estimated at 85 TWh [28]. The leading countries are the United States, the Philippines, Indonesia, Turkey, Mexico, New Zealand, Italy, Iceland, Kenya and Japan. They all have more than 500 MW in operation.

According to a study from 2013, there were 806 geothermal power projects in development around the world, with a combined capacity of 23,313 MW. The majority are located in Asia, North America and Africa [29].

Installations for geothermal heat generated about the same as the installed power worldwide, around 85 TWh. The capacity factor is substantially lower for heat extraction than for electricity generation. This is because production is controlled more by the need for heating and that is low in the summer.

Figure 2.44 does not include the energy output of geothermal ground-source heat pumps. They are very popular in many parts of the world. As an example, 250,000 units were sold in 2015 in the United States.

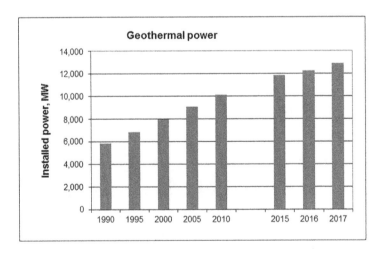

Figure 2.44 The capacity of installed geothermal power worldwide has grown steadily between 1990 and 2017 [27].

2.4.1. Renewable energy?

The geothermal resource is in part a remnant of Earth's formation 4.6 billion years ago. At that time the core temperature was even higher than today's core temperature of 4200° C. Heat leaks out constantly from the core towards the earth's surface and eventually radiates out into space.

It has been estimated that 42 TW from the earth's core is constantly radiating outwards. Of that, 20 TW comes from the earth's formation while 22 TW are derived from energy development resulting from the spontaneous radioactive decay occurring in the earth's crust. It is radioactive isotopes, mainly thorium 232, uranium 238 and potassium 40, which emit heat as they decay.

There is no doubt that both sources are limited and will decrease. But they decrease just as quickly, whether we use them or not, unlike fossil fuels, which are reduced if they are used, and the time frame for the decrease is long, so with a perspective of a few hundred years, the resources are usable.

But locally the temperature can drop in a geothermal source. That is especially true for a source of high temperature, so that the operation must end either permanently or for a period of recovery.

If 1% (0.42 TW) of the geothermal heat leakage could be used, it could provide 3.7 PWh per year, or 3% of the world's energy needs. Geothermal energy is a significant source of energy that can be considered renewable in our perspective.

2.4.2. Direct use of geothermal heat

People have since time immemorial been using geothermal heat. People started early to use hot springs for bathing and to heat and cook food. Today hot springs are often the basis for thermal baths and spa facilities.

In modern systems for direct use, a hole is often drilled into the underground reservoir to get an even flow of hot water. After the cooling, when the heat is exhausted, the water can either be returned in an injection well or placed in surface water.

Geothermal water can be used in many applications where heat is needed (Fig. 2.45). Sometimes water that only varies by a few degrees higher temperature than the surroundings can be used as really low-grade geothermal energy. Examples of such use can be in fish farms or for heat pumps.

Geothermal hot water can be used for heating individual houses, in district heating systems or to heat greenhouses for growing vegetables. Other applications that use heat are to dry grain, fruit, wood lumber and timber.

Only a moderate warming of the water in fish farms can provide a faster growth. Geothermal heat can also be used in various industrial processes for drying and heating, for example to pasteurize milk.

2.4.3. Geothermal power generation

Dry-steam power plants are the simplest type of geothermal power plants. They were the first type of geothermal power plant to be built. They use the steam from the source directly to drive the turbine. The first power plant was built in Larderello in Italy in 1904, and it is still in operation.

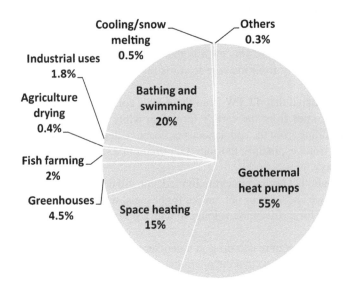

Figure 2.45 Geothermal heat can often be used directly. The percentages in the figure are based on the total geothermal heat used per year globally [30].

Dry-steam power plants are suitable if the source provides a superheated steam with temperatures of about 180–185 °C. Some of the water vapor is released to the surroundings but most of it is recycled in an injection source.

This power plant type is most used in Italy and at the geysers in northern California but also in Indonesia, Japan and Mexico. In California, the number of dry-steam power plants has increased in recent years and attracted many energy companies because of the economic potential. In some sources, the pressure has fallen in recent years, which can be an indication of over-exploitation. The efficiency of a dry-steam power plant can be about 30%.

Flash-steam power plants are the most common type of geothermal power plant operating today (Fig. 2.46). The water used has a temperature of more than 180 °C and must be pumped to the surface under high pressure. The water is sprayed into a tank of much lower pressure, where it is vaporized quickly – "flash". The steam then drives the turbine and the electric generator.

The remaining fluid can be supplied to a secondary tank with even lower pressure. From this, more steam and electricity are extracted. If there is heat demand, the remaining liquid can be used, for example, in a district heating network. Flash-steam power plants have a slightly lower efficiency than dry-steam power plants.

2.4.3.1. Binary cycle power plants

Most geothermal sources give flows with lower temperatures. Water with a temperature in the range of 100–170°C can be used in a binary cycle power plant (Fig. 2.47). It differs from other types of geothermal power plants in that the flow from the source does not contact the turbine directly. The power plants use a secondary liquid with a lower boiling point, such as butane. The liquid evaporates when it is heated by the geothermal flow in a heat exchanger and the steam drives the turbine.

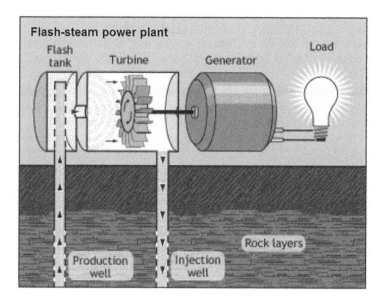

Figure 2.46 Flash-steam plants are the most common type of geothermal power plant in operation today.
Source: DOE/EIA [31]

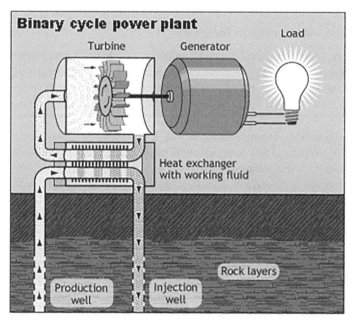

Figure 2.47 In a binary cycle power plant the heat from the rock is absorbed with a heat exchanger.
Source: DOE/EIA [31]

An advantage of this more complex system is that there are no emissions into the atmosphere, because the geothermal flow is occurring in a closed circuit.

The disadvantages are that the geothermal stream must be kept under pressure so that it does not evaporate; it requires large flows and about 30% of the resulting electrical energy is consumed in the process.

Although the achieved efficiency can be low – down to 10% – this type of geothermal power plant has the greatest potential for the future. This is because of the lower demands on temperature and purity of the source than a binary cycle plant needs.

Although it is possible to work with temperatures lower than 100 °C, the efficiency of the electricity output decreases.

Two types of cycles are used, Organic Rankine Cycle (most common) and Kalina cycle. The Rankine cycle is named after W. J. M. Rankine, a Scottish engineer and physicist. He was an important contributor to the science of thermodynamics. Rankine has developed a complete theory of the steam engine and indeed of all heat engines.

The Organic Rankine Cycle (ORC) technology is one way to convert heat to electricity. Its main applications are electricity generation from renewable heat sources (geothermal, biomass, solar), but it can also be used for heat recovery from industrial processes. ORC systems can range in size from kilowatts for domestic cogeneration to large geothermal power plants. According to a report in 2017 [32], more than 700 projects were identified, with a total power of 2.7 MW. A total of 76.5% of the plants use geothermal heat.

ORC uses an organic, high-molecular-mass fluid with a liquid-vapor phase change, or boiling point, occurring at a lower temperature than the water-steam phase change. The fluid allows ORC to recover heat from lower temperature sources that then is converted to useful work and finally to electricity.

2.4.4. Geothermal energy in Iceland

Iceland has very good conditions for geothermal energy. Iceland was formed by volcanic activity in the crack between two tectonic plates. It is estimated that there are 200 volcanoes in the country. About 30 are still active. The solid crust in Iceland is in some places only a few kilometers thick. Magma below the surface heats up the lava that is higher up and heats the groundwater.

Iceland has the most hot springs in the world. The country has identified more than 600 hot springs, 250 geothermal low-temperature fields with degrees up to 180 °C and 26 high-temperature fields with temperatures above 180 °C.

The largest hot spring in Iceland, Deildartunguhver, provides 150 liters of boiling water per second. In the south, the valley of Haukadalur has the biggest hot water source, Geysir, which gave its name to the phenomenon. At one time a 60-meter-high column of water was sprayed straight up in the sky, but today it is kept active with the help of humans.

Icelanders began early to exploit geothermal energy for heating and power generation. The first sources of hot water for heating were prepared in 1928 and 1930. The hot water was conducted in a 3-kilometer-long pipeline to Reykjavik and heated up a school. It was also joined to some office buildings and 60 houses. Geothermal heat is so cheap that even the sidewalks in Reykjavik and Akureyri are heated with it. Most district heating in Iceland comes from three geothermal power plants, which in addition to the electrical energy can deliver more than 800 MW of heat energy.

The total consumption of primary energy in 2017 in Iceland was 69.6 TWh (Fig. 2.48). A total of 61% came from geothermal sources, and with an additional 20% hydropower, the country has 81% of its energy demands met by renewable energy.

Nine large geothermal power plants produce 27% of the electrical energy (2017). The other electricity source is hydropower. According to statistics, Icelanders consume the most electricity per capita in the world. In 2017, they used 57,000 kWh per person. But 77% of electricity is used in heavy industries, for example aluminum smelters. If that part is counted off, the use of electrical energy is about the same as in other Nordic countries.

The needs of space heating are covered to 89% by direct geothermal heat. Additional heat is mainly covered with electrical energy (of which 27% is covered by geothermal energy). In addition to electricity generation and heating, geothermal energy in Iceland is also used to meet many other energy needs (Fig. 2.49).

Iceland's large resources of hydroelectric and geothermal energy could contribute to the energy supply even for the rest of Europe. For many years, there has been an idea to build an electric cable between Iceland and the British Isles. According to calculations, the cost of electricity from Iceland via a subsea connection between the UK and Iceland should be cheaper than from off-shore wind power. This new submarine cable, sometimes called IceLink, would be a high-voltage direct current cable (HVDC), with an output of between 700–1000 MW. It would be 1000–1500 kilometers long, which is longer than any existing submarine cable today. The longest undersea cable currently, NorNed, between the Netherlands and Norway is 580 km.

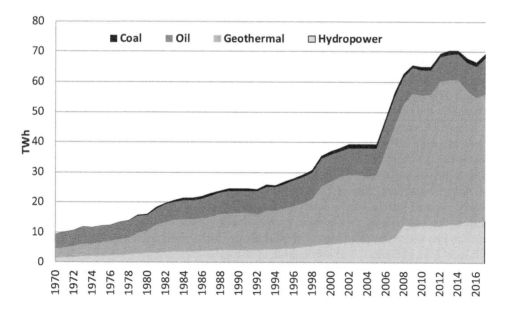

Figure 2.48 Primary energy use in Iceland in the years 1970–2017. Geothermal energy has increased dramatically and dominates today.

Source: National Energy Authority, Iceland [33].

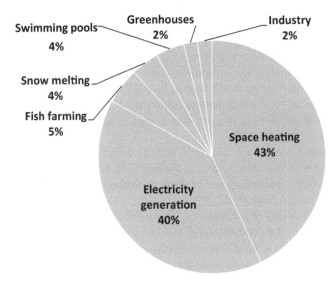

Figure 2.49 Geothermal energy in Iceland has a wide range of uses [34].

2.4.5. Small geothermal power plant

As small-scale geothermal power, projects under 5 MW of electrical power are usually included. GEOCAP (The Geothermal Capacity Building Program – Indonesia-Netherlands) is an international collaboration between Indonesian and Dutch interests with the goal of developing geothermal programs for education and research. On behalf of GEOCAP, the consulting company IF Technology has developed MiniGeo, which is a small-scale geothermal power plant for remote areas lacking electricity networks. The goal is to find an alternative to diesel generators, which is the most common option for these areas. These are both expensive and polluting.

The MiniGeo power plant is designed modularly and can be transported in a shipping container (Fig. 2.50). The advantages of MiniGeo compared with solar and wind power is that it can generate electricity 24 hours a day. It does not require any fuel, emits almost no carbon dioxide and needs only a small space. Devices producing between 100 kW and 1 MW of electric energy are planned.

The demand for electricity is not constant, so the MiniGeo system sometimes yields a surplus that can be used for other purposes. An idea is a desalination module that uses both heat and power to desalt seawater. Surplus energy can also be used in modules for drying products from agriculture and horticulture.

The calculated energy cost from a MiniGeo is higher than for electric energy from the electricity grid, but alternatives such as diesel generators are also more expensive than electricity from a power grid. The cost of electricity from a MiniGeo system is claimed to range from 0.10–0.20 US$/kWh depending on geological conditions and installation size. Power from diesel generators typically costs more than US$ 0.50 / kWh. This makes geothermal power an economically interesting option for remote communities.

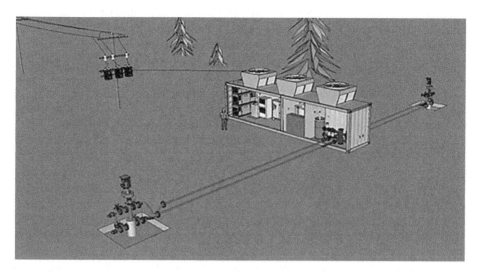

Figure 2.50 MiniGeo: a small, modular, geothermal power plant that can fit inside a shipping container. This pilot project can be constructed in Haruki Island, Maluku, Indonesia.

Source: IF Technology, Dutch energy consultant company [35].

The most important condition for installing MiniGeo is that the geothermal gradient in place enables relatively high temperatures at reasonable depth. According to the preliminary calculations, a gradient of 50 °C/km or more is in most cases sufficient. This gradient can be found in many places around the world.

Many places in Indonesia are suitable for this type of project. In the eastern part of the country there are thousands of remote volcanic islands where this type of power plant is suitable. The sites lack electrical grids but have large geothermal resources near the surface.

The first pilot project to be investigated is on the island of Haruku in the Moluccas region of eastern Indonesia. The island has a resting volcano in the middle and currently uses a 1.5-MW diesel generator to provide electricity to its 25,000 inhabitants. A MiniGeo system can reduce the need for the diesel generator and provide clean and stable electricity at lower costs.

Additionally they are looking into the production of drinking water and refrigeration as well as providing an internet connection.

2.4.6. Environmental impact

Geothermal energy has great advantages from an environmental perspective. A geothermal plant causes almost no greenhouse gases or other substances to be released into the air. If there are salts or impurities in the geothermal water, it is usually returned in an injection source. The energy can be recovered with full power around the clock, all year long. The visual impact is minimal. Only a small building is visible. That can be built so that it fits in well with the external environment.

An energy company in Lund, Sweden, uses geothermal heat for district heating. The company states that the heat produced in their geothermal system has approximately 30 times less impact on the greenhouse effect than if fossil fuel were used.

2.5. Bioenergy

Bioenergy we collect from growing plants – the biomass. It is the solar energy that has been converted into chemical energy through photosynthesis. Approximately 0.6 per mile of the solar energy that radiates to the earth becomes bound in biomass. It is about seven times the total human energy needs. Both in Sweden and globally, bioenergy is the largest renewable energy source. But it is still a limited resource. In many of the world's densely populated countries, bioenergy only contributes marginally to the energy supply.

Sweden is one of the countries that has the most favorable conditions for bioenergy, and it is also at the forefront of using it. Sweden used 143 TWh of bioenergy in 2017, which is 38% of the country's energy needs. It is a record-high level, which is made possible because our forests are growing faster than ever before. The cause is the fertilizer that we receive from nitrogen deposition (from the atmosphere, actually a consequence of fossil fuel combustion) and effective forest management.

Looking at the world, the situation for bioenergy is not always as positive. Not all countries have as much opportunity to harvest wood fuel and energy crops as does Sweden. But some countries with tropical climates can have a high growth of both crops (such as sugar cane and corn) and forest. All countries have the ability to extract energy from waste, such as in the form of biogas. However, there is often a lack of sustainable practices, for example regarding forestry and waste handling.

Using bioenergy for heating is an established technology, both small scale and large scale. Electrical power generation, often by combined heat and power (CHP, also called cogeneration), is also an established technology, especially for large-scale bioenergy. However, modern use of bioenergy for the most small-scale electricity generation has been developed more recently. According to one definition, small-scale-CHP is below 1 MW maximum electrical capacity and micro-CHP is below 50 kW [36], but other definitions can also be found. Small units have the potential for a higher total efficiency than traditional systems, because they can be located near the point of use and therefore losses can be avoided in the transfer of electricity and heat. Because electricity can be transported more practically, it is of special importance to locate CHP where the heat can be used. In some cases, in particular when using fuel cells, even the electrical efficiency can actually be higher in a small plant than in traditional systems (Table 2.7).

As energy systems can be expected to increasingly be supplied by intermittent sources like solar and wind, bioenergy can have an important role in the energy mix to supply dispatchable power throughout daily and annual cycles. Some energy carriers that are relevant for bioenergy are listed in Table 2.8 and shown in Figure 2.51.

2.5.1. Types of biomass

In Sweden, bioenergy is defined as "energy generated from biomass." The fuel may have undergone chemical, biological or mechanical conversion. Biomass is "material of biological

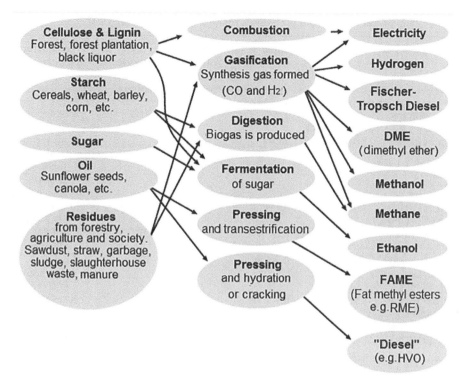

Figure 2.51 Energy carriers, conversion processes and raw materials for bioenergy.

Source: Based mainly on Maria Grahn, Physical Resource Theory, Chalmers University of Technology.

origin not or only slightly transformed chemically or biologically." Bioenergy can be roughly divided into five groups:

1. Wood fuels are wood raw material from the forest. These include forest residues (branches and tops or logging residues), small-diameter wood, bark, chips, sawdust and wood powder, pellets and briquettes.
2. Recovered liquors are a waste product of the pulp industry produced when woodchips are cooked to pulp. Liquors consist of a mixture of lignin, hemicellulose and the chemicals in the pulp cook. Today the liquors are fired in the pulp industry's recovery boilers, which also recycle chemicals.
3. Biofuels from waste: Solid waste is usually burned in incinerators. Liquid waste, such as sewage sludge, manure, food waste, slaughterhouse waste, etc., can be digested into biogas. Leaking gas from waste dumps, landfill gas, can be collected.
4. Peat fuel is retrieved from peat bogs. It's biomass that is incompletely decomposed and formed in bogs and marshes.
5. Agrarian biofuels come from agriculture. Energy grass such as reed canary grass, energy forest, often willow species, straw and cereals for incineration or the manufacture of ethanol are examples.

Table 2.7 Some of the smallest commercially available products that can be used for producing electricity from bioenergy, found at the time of writing

Technology	Fuel	Rated electrical output	Rated thermal output	Peak electrical efficiency approx.[*]	Manufacturer, product
Thermoelectric	Wood sticks, pellets, etc.	3 W		<1%	BioLite, CampStove 2
Thermoelectric	Propane, etc.	21 W			Gentherm Global Power Technologies, Global SolarHybrid
Direct methanol fuel cell (DMFC)	Methanol	45 W	130 W	25–30%	SFC Energy, EFOY Pro 800
Stirling engine	Wood pellets	0.6 kW	9 kW	10%	ÖkoFEN / Microgen
Proton exchange membrane fuel cell (PEMFC)	City gas / LPG	0.7 kW		39%	Toshiba / Panasonic, ENE-FARM
Solid oxide fuel cell (SOFC)	City gas / LPG	0.7 kW		52%	Aisin Seiki, ENE-FARM
Solid oxide fuel cell (SOFC)	Methane (upgraded biogas)	1.5 kW	0.6 kW	60%	SOLIDpower, BlueGEN / Energy-Company / SwillPower
Micro turbine	Methane (upgraded biogas)	3.2 kW	15.6 kW	16%	MTT, EnerTwin
Genset with internal combustion engine	e.g. HVO diesel	3.5 kW		26%	
Gasification and internal combustion engine	Woodchips	9 kW	22 kW	23%	Spanner Re², HKA 10
Organic Rankine Cycle (ORC)	Any	10 kW		7%	Enogia, ENO-10LT

* Based on lower heating value (LHV) of fuel in question.

2.5.2. Environmental impact

Biomass is considered to be a closed-loop cycle fuel, when done properly (Fig. 2.52). As with any combustion, carbon dioxide is emitted when biomass is burning. But the same amount of carbon dioxide has been taken up by the growing plants. Thus, no new carbon dioxide is released into the atmosphere.

The time of the cycle is less than half the average life expectancy of the trees, perhaps 20–40 years. For agrarian biofuels cycle time is less than one year. Thus, there is no net

Figure 2.52 Carbon dioxide from combustion is used in photosynthesis. Nutrients from the soil can be returned by the ashes.

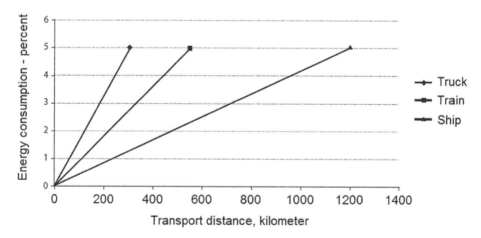

Figure 2.53 Example of energy consumption in the transport of woodchips. It is specified as a percentage of the energy content of the chips.

effect of carbon dioxide from bioenergy on the greenhouse effect (contrary to fossil fuels). However, the use of fertilizer must be kept under control so the much stronger greenhouse gas nitrous oxide is not formed.

Transport of biomass is commonly considered a problem. Fossil fuels are used in transport. But in some cases only 2–4% of energy content in the biofuel is needed for normal production and transport (Fig. 2.53). That is a higher proportion than for fossil fuels. In the future, even the energy for transportation can be collected from the bio sector.

Table 2.8 Some energy carriers that are relevant for bioenergy and comparison to fossil fuels, based on [37]. Higher heating value (gross calorific value) is based on that the water of combustion is entirely condensed and the heat contained in the water vapor is recovered (which actually depends on temperature). Lower heating value (net calorific value) is based on that the heat in the water vapor is not recovered.

Fuel	Higher heating value (HHV)				Lower heating value (LHV)				Comment
	MJ/kg	kWh/kg	MJ/L	kWh/L	MJ/kg	kWh/kg	MJ/L	kWh/L	
Gaseous (0 °C, 1 bar)									
Ammonia	22.5	6.25							
Hydrogen	141.7	39.4	0.0127	0.00353	120.0	33.3	0.0108	0.00300	
Methane	55.5	15.4	0.0398	0.0110	50.0	13.9	0.0358	0.00994	
Liquid (15 °C)									
Biodiesel (methyl ester)*	40.2	11.17	35.7	9.92	37.5	10.42	33.3	9.25	
Diesel*	45.6	12.67	38.6	10.7	42.6	11.83	36.0	10.0	Synthetic diesel similar
Dimethyl ether (DME)	31.7	8.81	21.1	5.86	28.9	8.03	19.2	5.33	
Ethanol	29.7	8.25	23.4	6.50	26.7	7.42	21.1	5.86	
LNG*	55.2	15.33	23.6	6.56	48.6	13.50	20.8	5.78	
LPG*	49.3	13.69	26.5	7.36	45.5	12.64	24.4	6.78	
Methanol	23.0	6.39	18.2	5.06	19.9	5.54	15.8	4.39	
Paraffin (wax)*	46.0	12.78	41.4	11.5	41.5	11.53	37.4	10.4	
Petrol (gasoline)*	46.4	12.89	34.2	9.50	43.4	12.06	32.0	8.89	

* Fuels which consist of a mixture of several different compounds may vary in quality between seasons, etc. The variation in quality may give heating values within a range 5–10% higher and lower than the given value.

Biological considerations are important to achieve sustainable practices, including preservation of biodiversity. Unwise use of biomass can have wide-ranging effects, for example on wildlife, including habitat loss.

The ground may not be used to the point that it is depleted. If the needles from conifers are left, the abduction of nutrients is limited significantly. If the tree branches and twigs are left on the site for drying and dropping needles in smaller piles, the nutrients are spread out over the cutting area. Then plants can better absorb the nutrients in the needles. Branches and tops carry a large part of the tree nutrients. By simply removing only a part of the branches and tops, the land can retain a large part of the nutrients.

Ashes from the combustion of biofuel contain nutrients abducted from the harvest area. Ashes should be returned so that the area's nutrients are not depleted in the long term. By ash recycling the nutrient balance can be restored and the loop can be close-circuited for many substances. Ash recycling may even produce a small increase in growth on some land. How the ashes are to be spread in the forest is a difficult question. The technology of the reversal is not fully developed yet.

In some cases ash recycling can help in the recovery of acidified ground. A reduced leakage of acid and aluminum-water to lakes and streams can be achieved. An improved quality of runoff water means better habitat for fish and other aquatic organisms.

There can also be emissions of particles, nitrogen oxides, etc., from burning of fuel. This can impart considerable negative health effects, especially when emissions take place near people, such as in individual homes in a city. In Sweden, cities often have local regulations limiting the ability to use polluting technology. There are established procedures for testing and labeling of products. For example, in Sweden "environmentally approved" means that the product has been tested and fulfills emission requirements. There are also other labels with requirements on emissions, efficiency, safety and operational aspects. In addition, there are usually requirements on regular inspections. Large-scale technology means better opportunities regarding treatment of flue gases, as well as moving emissions away from sites where many people are gathered.

2.5.3. Certification

There are different types of certification systems relevant to bioenergy. One example of the organizations working with sustainability of biomaterials is the Roundtable on Sustainable Biomaterials (RSB). Their certification scheme verifies that biomaterials are ethical, sustainable and credibly sourced.

2.5.4. Firewood

Firewood is the most common fuel for heating homes and cooking in the world. All of Sweden's housing was once heated with firewood.

If a wood-fired boiler is used, it is important to combine it with a heat storage system, usually in the form of a hot water accumulator tank, which will receive and store most of the energy released during the short time when wood burns with full power. This enables much more efficient use of the firewood compared with keeping a fire burning constantly. Firewood can also be burned in an open fireplace, a stove for cooking, a tiled stove or in other types of stoves. The classic open fireplace has the worst efficiency. It creates a large flow of air through the chimney that must then be replaced by cold outdoor air. Resulting heating output can be zero or negative. Something that encloses the fire and restricts the flow of air is a big improvement. A coziness factor can be retained if the stove insert has a glass door.

Hydronic heating systems are often used to spread the heat to multiple rooms. A fan can also be used to spread the heat. Soapstone stoves and tile stoves have an accumulating effect and therefore provide a smooth heat. The tiled stove is an old Swedish invention which was developed during the 1700s wood shortage. There are new tiled stoves on the market but often old ones are reused. New tiled stoves can have water jackets so that they can be connected to hydronic heating systems.

Small stoves and kitchen stoves are often used for cooking with firewood. For heating the disadvantages are that they normally only provide heat when the fire is burning. Kitchen boilers that heat water, for example for central heating, are also possible but relatively rare. Today in Sweden, small stoves and open fireplaces are used in some homes, vacation houses, cabin sheds and the like, or as a backup source, for example, when long-term power outages occur. In the developing world, cooking with firewood is more common.

Cooking with firewood is often done very inefficiently – a simple open fire can have an efficiency no better than 2% [38]. The thermoelectric generators available, for example to place on a firewood stove, can usually provide no more than a few watts of electricity.

One initiative to create a more efficient stove for cooking with firewood, especially in developing countries, is InStove. Their 60- and 100-liter models are specified to have a 50% average efficiency from startup to boil of half a pot, based on a standard test [39]. They also have equipment for sterilization of medical apparatuses and purifying water by burning firewood. Under development is an add-on thermoelectric generator.

Firewood requires much manual work, but is considered to be the one of the cheapest forms of energy, especially if you prepare the firewood yourself from your own forest and don't calculate the high cost for your own time.

2.5.4.1. Preparation of wood

The first step in preparation is wood harvesting. Birch is often the most advantageous wood because of its high energy content. Other hardwoods (oak, beech, maple, etc.) have higher energy content relative to weight, but they are limited in quantity.

Some wood is affected by rot, sometimes even when the trees are in the forest. The wood rot reduces the value for sawmills and pulp mills, but it is still useful for extracting energy (sometimes with only a marginal decrease in energy content).

Rot or mold, which can have negative health effects, can also occur during storage because of the high moisture content in wood, which also lowers combustion efficiency. This scenario can be avoided if the wood is cut and dried shortly after harvesting and if it is well aired during storage.

For the best quality, wood harvesting should be done in the winter. Then it is cut up and split, so that it dries during the best drying time, spring and early summer.

Today, there are many good tools for the production of firewood, such as splitters and sawing machines. They can either be powered by electricity or by a tractor. For larger users and wood producers, multifunction machines can handle sawing, cutting and transportation/packaging.

After preparation, it is important that the drying process start immediately and that the wood is airy and not too densely packed. Best is to dry the wood under a roof, but with open walls (Fig. 2.54). Covering the wood with a tarpaulin results in poor drying and the risk of

Figure 2.54 A wood storage shed with open sides and room for two years' consumption is ideal.

rot/mold. Drying wood takes time, say two years. When burning wood, the moisture content should not exceed 20%.

How much of the energy content becomes useful energy depends on furnace and operating technique. It is appropriate to mix wood from deciduous trees and conifer. A special case is oak. Oak wood contains much energy but also tannic acid, which can damage the furnace and flues. To avoid this, the oak wood can be co-fired with other wood. Table 2.9 shows the energy content when the wood is dried to 20% moisture content. If a villa's heating requires 25,000 kWh per year, the annual requirement is for about 16 cubic meters of birch firewood, stacked cubic dimensions. If the furnace is poor (still common today), the need increases to perhaps 29 cubic meters.

The efficiency of wood burning will be lower, if a long time passes between each fire. The boiler and tank will also emit a heat leakage in the house that can provide a bothersome over-temperature indoors during the summer. A well-dimensioned solar thermal system solves the problem, and also provides leave from work with wood handling and firing.

2.5.4.2. Environmental impact of burning wood

A major environmental advantage of biofuels is that it can be carbon neutral. But other emissions from combustion negatively affect the environment.

During complete combustion of wood the residues are carbon dioxide, water vapor and ashes. The smoke is white and odorless and the ash amount is small. Emissions that give health and environmental effects of burning wood are mainly from incomplete

Table 2.9 Energy content per solid cubic meter for different types of wood (which differs mainly because density is different) [40]. The figures refer to cut and split wood with 20% moisture content. Stacked cubic meter equals 60–70% of solid. Loose cubic meter equals 40–50% of solid.

Wood species		LHV per solid m³
Syringa (lilac)		16,300 MJ (4540 kWh)
Oak		13,000 MJ (3600 kWh)
	Some rot	10,700 MJ (2980 kWh)
	Much rot	9000 MJ (2500 kWh)
Apple		11,400 MJ (3180 kWh)
Rowan		11,200 MJ (3100 kWh)
Elm		11,000 MJ (3050 kWh)
	Some rot	8570 MJ (2380 kWh)
	Much rot	6190 MJ (1720 kWh)
Maple		10,800 MJ (3010 kWh)
Birch		10,200 MJ (2820 kWh)
Pine	Resinous	9290 MJ (2580 kWh)
	High density	8500 MJ (2360 kWh)
	House timber*	8280 MJ (2300 kWh)
	Low density	7200 MJ (2000 kWh)
	Soft rot	6620 MJ (1840 kWh)
Sallow		9000 MJ (2500 kWh)
Spruce	High density	8350 MJ (2320 kWh)
	Solid brown rot	8170 MJ (2270 kWh)
	Dry spruce	8030 MJ (2230 kWh)
	Low density	7490 MJ (2080 kWh)
	Soft rot	5760 MJ (1600 kWh)
Aspen		7880 MJ (2190 kWh)
Alder		7420 MJ (2060 kWh)

* About 150 years old. Thus, the energy content does not seem to deteriorate during prolonged storage.

combustion. Then carbon monoxide, heavy hydrocarbons (tar) and volatile organic compounds (VOCs, in practice half-burned gases, part of organic gaseous carbon) are emitted. Dust (fly ash) comes from soot at incomplete combustion and from non-combustible materials in the ash. Nitrogen oxides (NOx) are formed when the nitrogen in the fuel is oxidized. The formation can be reduced by control of the air supply and the construction of the boiler.

Risks associated with emissions from small-scale wood burning in boilers and stoves have attracted increasing attention. Small-scale wood burning is in some cases considered to be the largest single source of particulate emissions, followed by heavy diesel traffic and work machines. Emissions from small-scale burning of wood are estimated to be a significant environmental cancer risk.

Examples of health and environmental effects of the emissions are odor problems in the local environment, effects on the respiratory system for sensitive people and cancer risk from tars. Current research indicates that health effects of particles from combustion can be caused by some metal particles or high concentrations of very small particles.

The nitrogen oxides (NOx) contribute to eutrophication and acidification. The contribution to total emissions of NOx from combustion of firewood is low, however.

Sulfur emissions from burning wood are low. When replacing fossil fuels with wood, the total amount of sulfur emissions is reduced.

Volatile hydrocarbons can combine with sunlight to form ground-level ozone and photochemical oxidants. These can have health effects but also effects on crops, with crop losses as a result. In countries like Sweden, where wood burning occurs mainly in winter when the availability of sunlight is low, fuelwood contributions are limited.

The problems with emissions from burning wood are more related to outdated technology and improper technique than the fuel itself. If all outdated and inaccurate boilers were exchanged for modern, environmentally approved accumulator tanks, then harmful emissions would be reduced to a fraction. In Sweden, the installation of new boilers and stoves must meet the requirements of the National Board of Housing, Building and Planning. One way to comply with the rules is to use eco-labeled fireplaces. The Swedish Environmental Protection Agency recommends lighting a fire from the top, for reduced emissions.

It is important to use clean fuel – not waste, impregnated or painted wood, which may only be burned in facilities designed for this fuel and that have advanced purification equipment for the exhaust gases.

Dry wood and good technique provide high efficiency when burning wood, up to about 90%. For an older boiler without a storage tank, the efficiency can be as low as 50%. This means that about half of the energy goes through the chimney without any benefit. This is shown by the flue gas temperature, which may be 600 °C in an old boiler, but only 250 °C in a modern boiler that takes advantage of the heat.

By using upgraded equipment and training firewood users the environmental impact and efficiency of wood burning can be significantly improved and become fully acceptable long-term.

2.5.5. Woodchips

Woodchips are the simplest fuel for large-scale use of biomass. Woodchips are finely shredded wood treated in a wood-chipper. The raw material can be logging residues but also wood from forest clearing, pulpwood of bad quality or farmed willow. Even residues from the sawmill industry are used.

The raw materials can be chipped with small mobile chippers, which either have their own motor or are mounted on a tractor. Larger chippers can be leased, should larger amounts be needed.

Chips are used both for energy and as a raw material for the paper industry. As energy raw material it is used mainly in heating and power plants. Sometime chips are used in smaller facilities, but these chips must have a better, smoother and drier quality. Woodchips generally require more advanced equipment for burning than fuel pellets. The equipment has more moving parts.

The moisture content in woodchips varies, and is normally higher than in pellets and briquettes. The net calorific value increases with lower moisture content. The energy content of a cubic meter of woodchips with a 35% moisture content is around 900 kWh. For replacing one cubic meter of oil consumed, between 12–14 cubic meters of woodchips are needed. Since the chip is bulky, large storage areas are required. Woodchips should be stored under a roof.

During 2017 the price of woodchips in Sweden was about 0.18 SEK/kWh (0.018 €/kWh) (LHV) for large purchasers, according to Swedish Energy Agency statistics. No taxes are included in that price. Small consumers pay more. Prices have remained relatively unchanged in the past years.

The heating value is increased if the chips are dried and also depending on storability. Damp woodchips increase the risk of spontaneous combustion because of microbial activity and may contain harmful mold spores. Victims of mold spores can suffer from an allergic lung disease.

The disease expresses itself in the form of fever, chills, muscle aches and cough. The health risk makes woodchips not as appropriate in the family house sector, despite the low price.

2.5.6. Refined solid wood fuels: Pellets, briquettes and woodpowder

Pellets, briquettes and woodpowder are made from by-products from the timber industry, mainly sawdust. These refined products have a higher energy content than wood thanks to the standardized moisture level (Table 2.10). They also all have low ash content (<0.5%) and similar energy content by mass. Because of standardization and more consistent quality they are easier to burn with low emissions. The raw material is dried, milled and compressed under great pressure, to produce pellets and briquettes. They contain no additives. The wood's own adhesive – lignin – makes the material stick together.

Pellets (Fig. 2.55) are burned in pellet boilers, pellet stoves or pellet burners. The power from them can be easier controlled than with firewood, so there can be less need for an accumulator tank. Users are from the villa sector to large heating plants. Ordinary users are municipal buildings such as schools and sports centers, residential areas and industries.

It is easy to convert an oil-fired boiler to pellets. If one has an oil furnace in good condition it can be kept, and only the burner and equipment for feeding changed. The most developed pellet boilers have very high efficiency and low emissions of tar and unburned hydrocarbons. Boilers are available with automatic cleaning of the burner, sweeping of the convection surfaces and ejection of ash. Ash volumes become small, so the ash pan only needs to be emptied perhaps a few times a year. The boilers are comparable to burning oil in terms of maintenance needs.

A relatively new application of pellets is for small-scale CHP using a Stirling engine (see the case study "Self-sufficient home with pellet-driven CHP" in Chapter 6).

Pellets have a diameter of 6–12 millimeters and the length is a maximum of four times larger. A normal one-family house in Sweden consumes about 25,000 kWh of energy per year. With a pellet burner at 90% efficiency, 5–7 tons, or 10 cubic meters, of pellets are needed per year.

The price of pellets is about 50% higher than woodchips, if counted per kWh.

Briquettes are available in several forms. Briquettes can have a diameter or width and height of about 60–100 millimeters and varying lengths. Briquettes are mostly used in medium-sized boilers and range in size from 400 kW to 5 MW. However, sometimes in individual households briquettes are used in wood boilers.

Woodpowder is made from raw material that is dried and then grounded into finer particles according to user preference. It enables very efficient combustion, ensuring low emissions.

Table 2.10 Typical data for wood pellets and other fuels for comparison [41]. MC = Moisture content. Crown Copyright.

Fuel	LHV by mass		Bulk density	LHV by volume	
	MJ/kg	kWh/kg	kg/m³	MJ/m³	kWh/m³
Woodchips (30% MC)	12.5	3.5	250	3100	870
Log wood (stacked – air dry: 20% MC)	14.7	4.1	350–500	5200–7400	1400–2000
Wood (solid – oven dry)	19	5.3	400–600	7600–11,400	2100–3200
Wood pellets (10% MC)	17	4.8	650	11,000	3100
Heating oil	42.5	11.8	845	36,000	10,000

Source: Courtesy of Forestry Commission, licensed under the Open Government License.

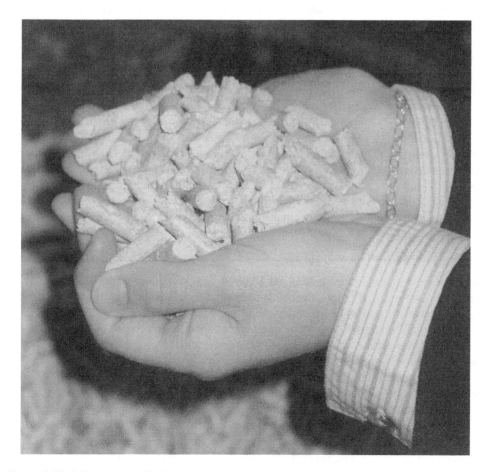

Figure 2.55 Pellets normally have a higher energy content than wood and woodchips because the moisture content is lower. Standardization also helps so pellets can be burned in fireplaces easily maintained with good environmental data.

This is a special product that is used by a few large users. Handling, transport and combustion takes place in closed systems.

Energy from woodpowder is mostly used in boilers that have a greater power than 4 MW. Old oil boilers can be converted to woodpowder. The burner must be replaced with a powder burner or dual fuel burner. Existing facilities can be used as storage but may need to be supplemented with a silo or other sealed container. Woodpowder has a low density, about 200 kg/m³, so a larger volume will be needed than for many other fuels.

2.5.7. Waste-to-energy

In Sweden, usual household waste is 70–80% of biological origin. As the proportion is so high, waste is usually considered to be renewable energy in Sweden. Energy can be extracted from waste in different ways. Incineration of waste, which is common for large CHP plants, is usually not suitable for small-scale applications.

Some types of organic waste are suitable for energy extraction through anaerobic digestion in biogas plants. This is discussed in more detail under "Biogas" later in the book.

One type of biogas is landfill gas, which is recovered from old landfills of waste. Landfill gas is formed spontaneously, as long as degradation of the organic material is proceeding. In Sweden, deposition of organic material was banned in 2005, so the amount of landfill gas is expected to decrease here in the future. By collecting landfill gas the increase in the greenhouse effect can be limited in two ways. Firstly we limit methane emissions, which are over 20 times more potent than carbon dioxide, and secondly we generate energy that can replace fossil fuels. VOCs in released landfill gas also contributes to the formation of photochemical smog.

Landfill gas can be of low quality, with a low content of methane, especially if the landfill is old. For this reason, traditional internal combustion gas engines (Otto cycle, spark-ignition) can be used for CHP on relatively new landfills (where the methane content is, say, above 40%) but some other technologies like external-combustion Stirling engines can work also on old landfills (with as little as 18% methane) [42].

Also sludge in wastewater treatment plants and industrial sewage produces biogas.

2.5.8. Energy from peat

The ice sheet over the Nordic melted away 10,000–15,000 years ago. In the Nordic, peat (and other soils) is formed after this. Peat is an accumulation of partially decayed vegetation or organic matters, formed under anaerobic conditions when soils become waterlogged or lakes are overgrown. Peat moss is one of the most common constituents.

Peat has been used as fuel for centuries. For example, peat was harvested in southern Sweden in the 1700s to solve the contemporary energy crisis. Peat fuel began to be used on a large scale in the 1800s, but has gradually been replaced by imported fossil fuels. In connection with the 1970s oil crisis the interest in peat as fuel reawakened.

Close to a quarter, about 10 million hectares, of the land area of Sweden is covered with peat. If the peat depth is 30 centimeters or more, the area is considered geologically as peat land. There are approximately 6.3 million hectares of peat lands in Sweden. According to investigations conducted in 1970, 350,000 hectares of the Swedish peat lands were appropriate for peat extraction. The surface used for peat extraction in Sweden today is less than 6500 hectares.

A peat bog grows in depth by an average of 0.4 millimeter per year in northern Sweden and 0.5 millimeter in southern Sweden. This corresponds to energy of about 20 TWh per year. Today about a fifth of our annual growth is harvested.

Extraction of peat is weather dependent. Dry and hot summers make it possible to collect a lot of peat, whereas low amounts of peat are collected in cold and wet weather. Because the annual collection rate is difficult to plan, it is necessary to build up buffer stocks which smooth out fluctuations. The peat fuel collected in Sweden is mainly used in some of the 40 Swedish heating and cogeneration plants that mix it with forest fuels.

Mixing other fuels with peat has been found to improve the efficiency of combustion and reduce maintenance costs. Problems with sintering, slag formation and corrosion in boilers, which often occur when burning biomass, are reduced.

Compared with wood fuels, peat has a higher content of sulfur and nitrogen, and in older plants this results in higher emissions of sulfur oxides and nitrogen oxides. With modern treatment technology, emissions from combustion of peat are comparable with emissions from other solid fuels.

Harvesting of peat affects natural and cultural values, and requires special permission in Sweden. When peat is harvested, the water level sinks and water runoff is changed. The pH value of the runoff water is lowered which may increase leaching of heavy metals. Runoff water also has an increased humus content, and the habitat is changed on the site.

One to two years before the peat is harvested, the vegetation is stripped and the bog drained. After harvesting, the peat is left on the site, where it is dried by the sun and wind. The moisture content is then reduced from about 90% to 45–50%.

Peat can also be processed. For example, the company Härjedalens Miljöbränsle AB (HMAB) in 1989 opened a factory in Sweden for manufacturing peat briquettes. The facility was originally built to supply district heating in the town of Uppsala. It has in recent years made briquettes with both peat and wood as raw material. Customers have been primarily plants in Uppsala and Stockholm. Transport is by railway.

The big question for peat is whether it should be classed as a renewable or fossil fuel. Views differ on that.

The peat that is burned as fuel has been formed over thousands of years. The cycle of other bioenergy that has a coal rotation period of a few decades does not apply for peat. When it is burned, carbon that has been bound a long time is released.

But can we not allow the burning of peat equivalent to our annual growth, which in Sweden is 20 TWh, a fivefold increase from today's use? The peat industry considers peat a biofuel as long as harvest is less than annual growth. In some cases, peat is classified as a "slow-renewable" fuel.

Another issue of climate change which is usually emphasized regarding peat is that harvesting of peat reduces the natural methane formation in the bog. Because methane is a much stronger greenhouse gas pollutant than carbon dioxide, the reduction would balance carbon dioxide during combustion, and the net effect is neutral or positive. However, reliable measurements and calculations of the balance are lacking.

2.5.9. Bioenergy from agricultural lands

Agricultural bioenergy potentially consists of:

*   Cultivated crops such as cereals, sugar beet and oilseed
*   Plants digested into biogas
*   Wood such as *Salix* (willow), hybrid poplar
*   By-products such as straw, dung and residues from the food chain

Figure 2.56 Salix (willow) is an energy crop that provides a high energy yield.

Through anaerobic digestion it is also possible to take advantage of residual products – manure, crop residues, organic waste – which also provide positive environmental effects in the form of reduced methane leakage and recycling of nutrients to cultivated land.

Calculations show that ethanol from wheat, CHP from *Salix* (willow) and diesel from oilseeds have the best profitability in Sweden. Other crops such as reed canary grass, hemp, poplar and hybrid aspen are not as competitive.

Salix has several advantages as an energy crop. It is perennial, gives a high yield per hectare and can directly be used in CHP plants (Fig. 2.56).

Competition in food production has attracted increased attention regarding using crops for biofuel production. For example, if using corn and wheat for ethanol, a shortage might occur and food prices might rise, which would affect the poorest people the hardest.

2.5.10. Ethanol

Bioethanol was the first biofuel to become established in the transportation market; it can also be used for electricity generation. It is normally used in spark-ignition internal combustion engines in the same way as petrol (gasoline). Almost all petrol sold in Sweden has 5% ethanol mixed in by volume. In vehicles intended for ethanol operation, the E85 is a blend of about 85% ethanol and 15% petrol (in Sweden during winter, the mixture sold as E85 is actually about 75% ethanol and 25% petrol, in order to improve cold-start properties at temperatures to –25 °C). In Brazil, almost pure ethanol (E95 or E100) is used as a fuel. Engine-driven gensets for ethanol are hard to find, but have been tested successfully with a modified engine [43]. Ethanol or ethanol-blended petrol with more than 10% ethanol should never be used in engines unless explicitly approved by the manufacturer.

It is also worth knowing that ethanol fuel blends may have a shorter shelf life than pure petrol (in particular alkylate, a synthetic petrol fuel of fossil origin suitable for small engines, has a long shelf life).

Ethanol has a higher octane rating than petrol. This can give the engine a slightly higher efficiency than with petrol. Ethanol has less energy content than petrol, so the consumption of E85 is usually 30% higher than for petrol.

In Sweden, there is also another ethanol-based fuel, called ED95, or Etamax D. It consists of about 95% ethanol, a small percentage of ignition improvers, denaturants, lubricants and anticorrosive additives in order to optimize combustion and operation in adapted compression-ignition (diesel) engines. The fuel is not compatible with E85, that is, you cannot use the ED95 in a flexible-fuel vehicle (designed to run on any mixture of petrol and ethanol, usually up to 85% ethanol).

ED95 engines run with the same high efficiency as ordinary diesel engines. At the same time, the clean burning of ethanol in an adapted diesel engine makes it possible to meet stringent emission requirements without a particle filter.

Ethanol-powered fuel cells have also been reported to be under development.

The net addition of the greenhouse gas carbon dioxide when running an engine on ethanol is 30–90% lower than when running on petrol, depending strongly on the raw material and process used to produce the ethanol. Table 2.11 lists the carbon dioxide emissions for the average quality of E85 and other fuels delivered in Sweden in 2017 (the numbers for ethanol could have been different, if ethanol had been produced in a different way). Emissions of health and environmentally hazardous substances from modern E85 cars do not differ significantly from petrol cars. The ethanol gives somewhat cleaner emissions than petrol because it does not contain sulfur or cyclic hydrocarbons, such as benzene. No soot is formed in an ethanol engine, because during incomplete combustion only acetaldehyde and a very small amount of formaldehyde are formed.

Emissions of nitrogen oxides and particulates from ethanol-powered buses are significantly lower than from diesel buses.

Unlike conventional diesel and petrol, ethanol contains no carcinogenic substances and is completely degradable in the event of spillage.

There has been some skepticism about ethanol as a fuel. This is in part because ethanol production competes with food production. It also is because ethanol production is energy intensive, requiring many processes that need much energy – sometimes more energy than ethanol provides. There is new legislation regarding this in the EU.

It is generally considered that ethanol does not have sufficient potential to solve all the energy needs for today's traffic. No one fuel is the solution for everything. Solutions will perhaps vary in different parts of the world. For instance, Brazilian ethanol can now compete in price with fossil fuels without subsidies.

Table 2.11 Average carbon dioxide emissions in Sweden in 2017 from different fuels, including comparisons to electric car [44]

	g CO_2 equivalents per MJ (LHV)	g CO_2 equivalents per kWh (LHV)	Energy use kWh/km for average car 2014	g CO_2 equivalents per km for car
Diesel	79.3	285	0.55	157
Petrol	90.8	327	0.73	239
FAME	31.1	112	0.55	62
Biogas/CNG*	18.9	68	0.64	44
E85	48.8	176	0.69	121
HVO	11.1	40	0.55	22
LNG/LBG	73.9	266		
ED95	28.4	102		
Electricity**	13	47	0.15	7

* Based on the average vehicle gas mix in Sweden in 2017, which was 90% biogas and 10% natural gas.

** Based on Swedish electricity mix.

2.5.11. Biogas

When organic material is decomposed by microorganisms in an oxygen-free environment, biogas is formed, which consists primarily of methane (usually 50–75%) and carbon dioxide (Fig. 2.57). The gas also contains low levels of hydrogen sulfide, nitrogen and water vapor. Biogas is produced by anaerobic digestion of, for example, manure from agriculture, sewage sludge, food industry waste or sorted organic household waste. Many wastewater treatment plants today produce their own biogas to cover their energy needs. Biogas leaking from waste landfills (landfill gas) is often of low quality and can be collected, as already described. Digester gas and marsh gas are other names for biogas.

Biogas is an energy carrier that can be used for many purposes. Examples are combustion in a boiler, engine operation for electricity generation or vehicle operation.

Spark-ignition engines can run on almost untreated biogas (raw gas), for example in small-scale CHP plants. But there can be problems with the operation, including that hydrogen sulfide presents a corrosion risk.

By purification from hydrogen sulfide, carbon dioxide and moisture, a better fuel is obtained that meets the Compressed Natural Gas (CNG) fuel specification for cars, with about 97% methane. Such upgraded biogas, sometimes called biomethane, can also be fed into the natural gas network.

A biogas/CNG car has basically a standard spark-ignition (petrol) engine, with a separate fuel system for gaseous fuel. Cars that can run on both CNG and petrol are usually called bi-fuel cars. They are equipped with both a petrol tank and gas cylinders. Often petrol is needed for cold starts and the car then automatically switches over to CNG, without interruption in the driving.

An engine operating on biogas/CNG produces significantly lower levels of nitrogen oxides, hydrocarbons, carbon monoxide and soot particles in the exhaust gases than diesel and petrol. The gas can be compressed and stored in steel cylinders. When it is used in

Figure 2.57 Farm-size biogas plant at Plönninge, an agricultural school in Halland, Sweden. The facility consists of a digester, a secondary digester and a house for the operation.

vehicles, the gas is usually compressed to at least 200 bars (still gaseous) to limit the gas cylinder volume. At compression about 10% of the energy content is consumed. The refueling must be done with gas-tight connections, but it is not more difficult than for other fuels and time for filling is also approximately the same.

A mixture of CNG and hydrogen has been reported to be advantageous regarding emissions and efficiency, but may require gas storage adaptations and engine optimizations.

There is also a development to use biogas for fuel cells, which should provide lower emissions and higher electrical efficiency than engine-driven gensets. Examples of CHP units using solid oxide fuel cells (SOFCs) are Convion in Finland and BlueGEN/Swill Digester of SOLIDpower/Energy Company in Germany/Netherlands.

Biogas/CNG is (like petrol) flammable and can even be explosive under certain conditions. Storing gas under high pressure can also pose hazards, and the gas cylinders have a limited lifetime (usually 15–20 years). In Sweden, cars using biogas/CNG must go through a special regular inspection, where the gas cylinders are also made visible, so possible damage (such as corrosion), fastening and tightness can be inspected. On the other hand, testing of exhaust gas emissions can be simplified for biogas/CNG cars, because the exhaust is so clean.

Gas can also be cooled to a liquid, which reduces volume. From upgraded biogas a product called liquefied biogas (LBG), or liquefied biomethane (LBM), is created, which can more easily be transported (via insulated tanker ships or trucks designed for transportation of cryogenic liquids) and can be dispensed to either LNG vehicles/gensets or CNG vehicles/gensets.

The residual product, the digested substrate, is a valuable fertilizer. When the substrate is digested, plant nutrients are mineralized, so it can more easily be absorbed by growing plants. But if biogas plants are expanded comprehensively, the areas for spreading of the residue, called digestate, could become a limiting factor.

As usual, what you put in you will get out. For land used in organic farming there are requirements on what the substrate is allowed to contain in order to use the digestate. To increase acceptance of bio-fertilizer, and meet various food organizations' requirements for spreading on cultivated land, the bio-fertilizer can become product-certified. The certification involves requirements regarding raw materials, suppliers, collection and transport, reception, treatment process, product and instructions. With this certification, the digestate is turned into a quality-assured product.

As for other bioenergy, the carbon dioxide released during combustion of biogas has previously been bound by plants. But the methane itself is a much stronger greenhouse gas than carbon dioxide. Leaking methane contributes more than 20 times greater to global warming than the same amount of carbon dioxide. It is therefore important to avoid leakage from digestion, upgrade and use.

Methane concentration in the atmosphere rose significantly when in the 1700s the Chinese increased rice cultivation on flooded lands. Also, large-scale livestock farming contributes to the amount of methane in the atmosphere. All flooded land surfaces where there is organic matter, peat bogs, manure pits, dumps and water treatment ponds emit methane gas. Hot climates produce especially large amounts of methane.

A large portion of the substrate used for biogas production would still give rise to leaking methane, if it were not used for biogas. Good biogas facilities should be able to lower total methane emissions to the atmosphere and thereby reduce climate impact.

2.5.12. Biodiesel – FAME

The product called biodiesel is usually fatty acid methyl ester (FAME). The most common raw material is rapeseed oil transesterified to rapeseed methyl ester, RME. Rapeseed is an oil plant cultivated in many parts of the world, to some extent also in Sweden. Palm oil, soybean oil, maize oil and sunflower oil are other vegetable fats used for the production of FAME. Even animal fats can be used, as well as reusing frying oils, for example from hamburger bars.

FAME is mainly used in low blends in diesel fuel so that it is not necessary to make any adjustments to the engines. According to European standard EN 590, it is possible to mix up to 7% FAME in diesel fuel (without labeling of bio content at retail points).

To run the engine on pure FAME, the vehicle manufacturer must approve its use. EN 14214 and ASTM D6751 are two standards that apply to FAME. FAME can be used in conventional diesel engines adapted for it; FAME is somewhat more corrosive than regular diesel fuel.

Pure FAME operation means that the carbon dioxide emissions are typically reduced 50–80% compared with conventional diesel operation. Emissions of particles are usually slightly lower when using FAME, while emissions of nitrogen oxides are at the same level or slightly higher. Engines must be optimized for FAME – otherwise emissions can be higher. Biodiesel is biodegradable and non-toxic.

The water affinity of FAME can potentially cause problems. FAME is hygroscopic, which means it absorbs water from the atmosphere or its surroundings. Microbes can grow in the water. It is recommended that FAME not be stored longer than six months or one year because of the risk of microbe formation. This also applies to diesel with a low proportion of FAME.

When blends of biodiesel and conventional diesel are distributed for use in the retail diesel fuel marketplace, a system known as the "B" factor is often used to state the amount of biodiesel in any fuel mix. For example, B20 is 20% biodiesel and 80% conventional diesel.

Tests have also been done to run diesel engines with pressed rapeseed oil. Some farmers drive tractors on self-pressed rapeseed oil, and this has been done with diesel gensets too. However, normally the rapeseed oil must be preheated, and both the start and stop procedure is carried out with conventional diesel fuel. The Elsbett motor can be run on pure rapeseed oil, which can be made on a farm, and does not always need preheating.

2.5.13. Gasification and wood gas

Gasification is a process that at high temperature converts biomass (or fossil fuel) to carbon monoxide, hydrogen and often some carbon dioxide. The output gas is called synthesis gas (syngas), producer gas or pyrolysis gas and can be used for a variety of chemical processes, such as producing liquid fuels as described in the following sections, and for direct power production (see the case study "Farm with woodchip-driven CHP and solar PV" in Chapter 6), potentially with higher efficiency than direct combustion of the original fuel.

Gasification can take place either by pyrolysis (dry/destructive distillation by reacting the material in an oxygen-free environment at high temperatures, say >700 °C) or partial combustion (with a controlled amount of oxygen and/or steam). During partial combustion, the output can also contain nitrogen gas from the supplied air, which can be avoided by using pure oxygen.

Wood gas arises from partial combustion of wood and is closely related to synthesis gas. Normally, the wood gas also contains pollutants, primarily tars, which should be cleaned to, for example, avoid availability problems. Carbon monoxide and hydrogen gas are combustible, for example in an internal combustion engine. A car running on wood gas uses about 3–5 kg of wood to replace one liter of petrol.

A risk is that carbon monoxide is highly toxic.

The output will also have a solid part (char), which can be incinerated to provide heat to the process. A Swedish company, Cortus Energy, is developing technology to instead burn the dirty pyrolysis gas to generate heat to the process and gasify the clean char with steam, which they claim results in a zero-tar clean gas.

2.5.14. Synthetic diesel – HVO

Hydrogenated vegetable oil (HVO) is one type of synthetic diesel fuel. It is the biofuel which has increased most rapidly in recent years in Sweden. HVO is also the fuel that presently provides the greatest reductions of greenhouse gas emissions in Sweden, more than 80% (compared with fossil fuel). However, when palm fatty acid distillate (PFAD), a by-product from palm oil production, is used for HVO, concerns have been raised regarding further expansion of plantations into fragile tropical forest areas and peatland.

HVO is prepared by hydrogen treatment of vegetable oil or animal fat. By the chemical composition, it is a hydrocarbon that is comparable to conventional diesel fuel. It can be used as such in all modern diesel engines or also blended with fossil diesel. There is no demand for additional investments to use it. However, the engine warranty may not be valid, unless the engine is approved for HVO (specified by the standard EN 15940 for paraffinic diesel fuels).

Two other types of synthetic diesel are gas-to-liquid (GTL) and biomass-to-liquid (BTL), which can be produced by a Fischer-Tropsch process and have similar fuel properties as HVO. All these are synthetic paraffinic fuels (unlike FAME, which is an ester produced in a different way).

An example of a commercially available GTL fuel is EcoPar A, which is an ultra-clean diesel fuel fulfilling the ordinary diesel standard EN 590, produced from methane from natural gas, but the same process could be used with biomethane. The EcoPar company is also working on a development project on BTL from fully renewable sources.

The classic synthetic diesel fuel is Fischer-Tropsch (FT) diesel. It can be obtained from many different raw materials, both renewable and fossil. The Fischer-Tropsch process was invented by German researchers Franz Fischer and Hans Tropsch in the 1920s. During the World War II oil shortage in Germany, the process was used to produce synthetic diesel fuel. After the war, interest fell because of low crude oil prices.

In South Africa, Fischer-Tropsch plants were built to supply the country with petroleum products when it was cut off from oil imports because of the apartheid policy. South Africa has a good supply of coal. The plants remain and have been extended even after the boycott was stopped. They produce petrol, jet fuel and diesel oil, and a variety of different petroleum products.

Synthetic diesel can achieve considerable reduction of exhaust gas emissions compared with conventional diesel. Regarding storage and cold starts, synthetic fuels can have better properties than conventional diesel. Synthetic diesel can also be made non-toxic and biodegradable, in case of spills.

2.5.15. Methanol

Methanol, the simplest of the alcohols, is an important basic raw material in the chemical industry. It can (like ethanol) be used as an engine fuel in pure form or by blending with petrol. Direct methanol fuel cells typically in the size up to a few hundred watts is an established product, available, for example from the German company SFC Energy (see the case study "Hybrid system with direct methanol fuel cells" in Chapter 6). Larger methanol fuel cells usually have a reformer, for example Serenergy fuel cells.

Methanol was previously prepared by heating the sulfuric acid–treated wood in an oxygen-free environment, so-called dry distillation. (It was then also called wood alcohol, a term that should be avoided. The risk of confusion of the name with drinkable types of alcohol has certainly contributed to a number of cases of poisoning.) Methanol can cause serious poisoning in humans because of how it breaks down in the body. Methanol provides an indirect influence in that it is broken down in the liver. Formic acid and formaldehyde are formed and blood pH drops to a dangerously low level. The formic acid is suspected to be the substance that harms the eyes and causes blindness.

Industrial methanol production today is based on natural gas; there the methane by so-called shift, a chemical process in the presence of certain catalysts, is converted into methanol.

By gasification of the wood material, a form of destructive distillation, approximately 95% of the energy is converted to a gaseous mixture of hydrogen, methane and carbon oxides (CO and CO_2). From the gas mixture, methanol can be obtained by shifting. During the process, approximately 55% of the wood's energy content can be converted to methanol and 45% to heat. A plant should therefore be located where there is a demand, for example for district heating.

The energy content of methanol is less than half compared with petrol. It compensates for this by having a high octane number, however, meaning compression in the engine can be raised and thus efficiency can be improved.

Methanol is a high-octane motor fuel that is regularly used in motorsports as fuel in speed-way motorcycles and in certain racing cars, especially in the USA. One reason for this is that methanol is not as flammable as petrol and any fire can be extinguished with water.

Methanol can be produced by energy forests at a cost that is significantly lower than for ethanol produced by fermentation of agricultural products. Methanol from forest raw material provides a considerably greater energy yield than ethanol.

Methanol can also be corrosive to metals, gaskets and seals. Another characteristic is that it is hygroscopic.

2.5.16. Dimethyl ether – DME

Dimethyl ether (DME) is a gaseous fuel, which with good efficiency can be used in compression ignition (diesel) engines with a modified fuel system and is also a possibility for fuel cells. At normal temperature and pressure DME is a gas, but it turns into a fluid at relatively low pressures and has properties similar to liquefied petroleum gas. For it to become fluid the tank must be pressurized to 5 bars. In terms of volume a DME tank becomes approximately twice that of a diesel tank for the same energy content.

DME can be produced directly from synthesis gas through a catalytic process, but can also be manufactured by so-called dehydrogenation of methanol. The fuel can be manufactured either from non-renewable raw materials such as natural gas, or from biologically renewable resources such as forestry residues. DME is produced worldwide for purposes such as propellant in spray cans. Also DME-powered gensets exist. In addition, DME can power certain fuel cells, either directly or through a reformer.

DME from biologically renewable raw materials (biomass) is called bio-DME. Regardless of feedstock, DME-powered engines provide great energy efficiency, sootless combustion and low emissions of particulates and nitrogen oxides (NOx). Bio-DME also provides up to a 95% lower climate impact than conventional diesel fuel.

During normal storage, DME will not deteriorate over time. In case of a spill, DME evaporates, but does not deplete ozone. DME is highly flammable but relatively non-toxic.

2.5.17. Biopropane/BioLPG

Liquefied petroleum gas (LPG) is traditionally made from natural gas processing and petroleum refining. LPG is a group of flammable hydrocarbon gases and can contain propane, butane and isobutane (i-butane), as well as mixtures of these gases. All of them can be compressed into liquid at relatively low pressures. There is a special danger with LPG, because (unlike natural gas) it is heavier than air. Suffocation can occur as a result of LPG displacing air, especially if one becomes unconscious in the absence of oxygen. There is also a risk of explosion if the mixture of LPG and air is within explosive limits.

In 2018, it was reported that Neste started up large-scale renewable propane production in Rotterdam. The refinery there primarily produces Neste's renewable diesel. Now it also purifies and separates renewable propane from the sidestream gases produced by the refinery. Others are investigating other paths to produce biopropane.

2.5.18. Emerging technology

Algae and aquatic plants grown in lakes and in the sea can become a great opportunity for bioenergy. Algae can be used as raw material for biodiesel, bioethanol and hydrogen, and can even be used directly for combustion. Technology for utilization is not developed yet, but many research projects are under way around the world.

It is also likely that fuel cells will to a greater extent be possible for use as bioenergy. For example, researchers at Linköping University have developed a fuel cell that runs on lignin [45]. A tree consists of 25% lignin. Lignin is a cheap by-product from pulp manufacturing.

Another area of development is other types of synthetic fuels from renewable sources, such as synthetic petrol from sugar.

2.5.19. Energy consumption during production of biofuels

The energy needed to produce, distribute and fill up with biofuel varies widely for different fuels, raw materials and production methods. Generally, more energy is used to produce biofuels than what is required for the equivalent amount of fossil fuels. The reason is that the fossil oil is already floating, and extracted in large quantities where found. The energy consumed in refineries to produce petrol and other fossil fuels is a fairly small part of the total fuel energy content.

Natural gas, which can also be used in vehicles, must sometimes be purified but need not be refined. The distribution pipelines are not as energy intensive. However, for longer distances, pumping stations or converting gas to liquid by cooling (LNG) and transports with tanker are required. Filling up, for example a vehicle also requires an energy-intensive compression of the gas.

The total amount of energy required to extract, refine, distribute and fill up with diesel and petrol normally is 10–20% of the fuel's energy content. The variation for natural gas is greater and can in the worst case (LNG) be considerably higher than for petrol.

When biological raw materials are used to produce liquid fuels, it is often from solid raw material. It has to be converted to liquid form, or gas and then optionally refined. This process is more energy consuming compared with the need for fossil fuels. Typically at least 50% of the energy content of the finished fuel is needed for production and distribution, but it can be either higher or lower proportions.

If the waste heat in the processes can be utilized, one can get a higher total efficiency, but the increasing production of biofuels, combined with better insulated buildings in the future, may make the heat difficult to utilize. The largest losses are obtained on raw materials, for example, ethanol, rapeseed oil/RME or biogas produced by intensive agriculture and converted in smaller facilities with low energy efficiency located far from the end users. Large-scale gasification of wood raw material for producing, for example, DME, methanol or synthetic diesel (Fischer-Tropsch diesel), provides the highest energy efficiency. The technology is far from fully developed.

As the availability of biofuels is limited, both production and the use of biofuels must be effective, if they are to replace the bulk of fossil fuels. Major research and development efforts will be required, but may still not result in the ability to replace all fossil fuel.

2.6. Engine-driven generators

Engine-driven generators, also called generating sets or gensets, are available with different kinds of engines, for different fuels and in a wide variety of sizes. This section focuses mainly on diesel gensets, because in off-grid renewable energy systems, they are common as a backup or complement to other forms of generation. Diesel engines are compression-ignition engines and traditionally use fossil diesel fuel, but modern fuels like HVO diesel could also be from renewable sources as explained in the previous section.

An obvious difference from some forms of renewable energy, such as solar and wind, is that gensets can produce power continuously, as long as there is fuel. Another difference is that gensets will – in addition to the electric generation – produce much heat. Under some circumstances, especially in systems where a genset is always in operation, it may be of interest to use this heat for something useful, for example space heating, tap-water heating or desalination of seawater.

It is also worth noting that gensets typically have a low capital cost but high operation and maintenance costs (mainly because of the fuel consumed), which is the opposite of, for example, solar, wind or hydropower, which require a comparatively large investment but have low operation and maintenance costs (their "fuel" is free).

2.6.1. Genset basics

Conventional gensets consist of an internal-combustion engine driving a generator at a (nearly) constant speed, which produces utility-grade AC, and are made to regulate voltage and frequency. Often 1500 rpm is used for gensets for 50 Hz and 1800 rpm for 60 Hz. Diesel is the dominating fuel for conventional gensets and normally provides better fuel economy than petrol (gasoline) gensets with a spark-ignition engine. Sometimes, gaseous fuels can be use instead (often in a diesel engine that has been "converted" to spark-ignition [46]).

Small gensets are usually built to be connected, one at a time, directly to a load or to an isolated grid. Control equipment to synchronize and load share with other power plants is usually optional, and is seldom seen on gensets below approximately 150 kW-rated power. MW-size power stations normally use multiple diesel gensets, which can synchronize and load-share with each other (Fig. 2.58a,b).

Some gensets have more than one power rating. Often there is a standby rating, which is for supplying power for a limited time (no sustained overload capability), and a prime rating, for unlimited running time (and 10% overload capability a limited time).

Traditional gensets have several environmental disadvantages. Air pollution and noise are some of the most obvious. The fossil fuels normally used today are consumed at a much higher rate than they are formed. Therefore, they are not designated renewable and contribute to the so-called greenhouse effect. Possible fuel or lube oil leaks, handling of waste lube oil and starting batteries are some other environmental concerns.

Regarding safety, carbon monoxide poisoning can cause death if a genset is operated indoors. Exhaust and genset compartment can also get very hot.

The characteristics of conventional gensets make them very interesting for applications such as emergency power, where the gensets only operate occasionally. However, for some applications, the starting time of the gensets must be considered. Therefore, gensets are in some cases used in combination with one or several uninterruptible power supply (UPS) systems, which can provide battery power to high-priority loads while the genset(s) starts up. Battery-based renewable energy systems can have a similar functionality. An alternative is to use a "no-break"-type genset, equipped with flywheel and clutch, which enables an instantaneous start without power interruption. However, such gensets are not common and cost more than the ordinary type.

Figure 2.58a,b Examples of diesel gensets of different sizes. Top: 3-MW genset in power station in Kotzebue, Alaska (where there are several MW-size gensets, which can operate in parallel, and the town is also supplied to a large extent by wind). Bottom: 12-kW genset at a remote tourist lodge in Amazonas, Brazil.

Source: S. Ruin.

On the other hand, when gensets are used continuously for prime power production, the disadvantages (high operation and maintenance costs, environmental issues and dependence on fuel, which is usually fossil and not produced at the site) are typically more noticeable.

(For applications where electrical power is not needed, engines are normally connected directly to the load, such as a water pump.)

2.6.2. Part-load performance

For many applications, one of the most important characteristics to be aware of is the part-load performance of gensets. For example, when renewable energy is introduced in a diesel grid, the average load of the diesel genset(s) will decrease (unless e.g. energy storage is used to prevent this). To a certain point, this is straightforward. Speed governors of gensets are usually fast enough to accommodate renewable energy, up to a certain point. Reverse-power conditions for the genset must normally be avoided. There are two more challenges associated with low-load operation, described next.

The first is efficiency. One can imagine that for an ideal genset, fuel consumption would go linearly to zero if the load on the genset goes to zero. However, reality is far from that if the genset will operate under low-load conditions. A conventional genset will have significant fuel consumption even at no load (can be around 20–30% of full load consumption for a small genset). Thus, genset efficiency is usually very poor at low load, as shown in Figure 2.59 for three small gensets. In this size range, the relative efficiency vs. relative load is rather similar, at least for these three gensets (Fig. 2.60). Larger gensets have better efficiency (to about 50%) and the fuel consumption vs. load is closer to an ideal genset, but to oversize the genset is counterproductive. A power station with multiple gensets can achieve high efficiency over a wide load range, so the total will be closer to an ideal genset.

The second problem is that low-load operation can cause maintenance issues and unplanned downtime. It can also negatively affect after-treatment components, such as diesel particulate

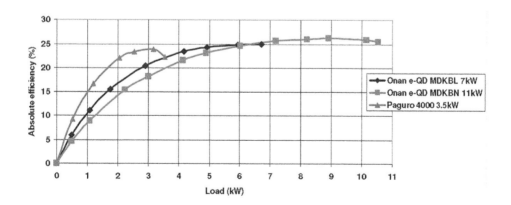

Figure 2.59 Absolute efficiency (based on the higher heating value 45.6 MJ/kg of diesel fuel) of three conventional gensets, which are representative for their size, according to independent tests by Victron Energy [47]. The power ratings are for continuous power.

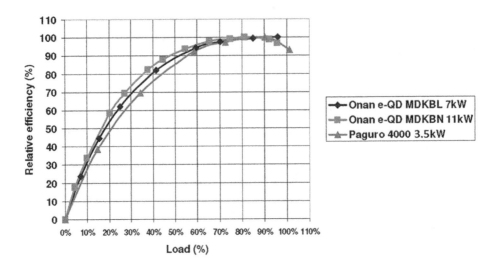

Figure 2.60 The test results by Victron Energy, shown as relative efficiency vs. relative load [47].

filters. Many gensets have a minimum recommended load to avoid such problems. This minimum genset loading can be as high as 30–40% of rated power, and significantly limits the amount of renewable energy that can be used. The genset should only be operated for a limited time below the specified minimum load.

Luckily, these problems have been reduced on some gensets. There are some special low-load capable gensets on the market. One approach is to run the genset with variable speed, which has many benefits. Through an inverter, the power output to load is constant frequency, regardless of the speed of the engine. The power quality of these inverters is often better than from other types of small generators.

2.6.3. Challenges with small gensets

Small gensets often have very simple governors, which can affect the output. For example, the 12-kW genset in Figure 2.58b could be set to a lower operating speed than the nominal 1800 rpm, and therefore sometimes be operated with an output of only 70 V AC (single phase) at 44 Hz, despite 60 Hz being the standard frequency in Brazil.

The following are some observations from tests by Victron Energy on gensets up to 11 kW:

- Although all gensets tested had a sound enclosure, sound levels varied more than expected. Northern Lights M773LW2.7, a 7-kW diesel genset with 1500 rpm, had the lowest sound level. Mastervolt 3.5 Whisper, a 3-kW diesel genset with 3000 rpm, had the highest.
- The report mentions that 1500 rpm gensets have a service life of up to 10,000 hours, while 3000-rpm gensets have up to 5000 hours, but that was not tested.
- Output frequency could vary almost 10% from nominal (50 Hz in this case), and usually – but not always – drops with increasing load. Victron's MultiPlus and Quattro

inverter/chargers operate in parallel with nearly all generators tested. The best results were obtained with generators with brushless, self-excited, externally voltage regulated, synchronous alternators (synchronous AVR) and "inverter" generators.

- Instead of running the generator for long periods under 30% load, the genset can be used together with an inverter/charger/battery system (so the genset can operate more efficiently and be stopped at low load). This leads to substantial reduction in fuel consumption, even if charging and discharging can generate losses from 25 to 35% for the energy that flows through the inverter/charger/battery. NOx and CO emission reduction will be even more impressive than the reduction of fuel consumption and the related reduction of CO_2.
- Down-sizing generators (e.g. by an inverter/charger/battery system) will increase the average load. This will reduce maintenance to an unknown extent, and will increase life expectancy.

When the Swedish Energy Agency in 2015 tested eight small gensets with rated power up to 5 kW, they found several shortcomings [48]. Only three of them lasted for the full test of 150 hours of operation. The power quality was for some types not good enough to power sensitive loads. Two of the tested gensets had emissions above the acceptable limits. Most of the tested gensets could not be started at −18 °C. Of the tested gensets, only Honda EU30 performed well in all parts of the test. It is a petrol-driven, 3 kW-genset with variable speed and inverter output.

According to a study on diesel generators in India by Shakti Sustainable Energy Foundation and ICF International, there is potential to improve small gensets sold in India. One of their recommendations is launch a standard and labeling scheme for diesel gensets [49].

2.6.4. Possible future steam gensets

A Swedish company named Ranotor envisions that modern steam engine technology has great potential for small-scale electricity generation [50].

The steam engine has continuous external combustion and highest efficiency at part-load. Like the Stirling engine, it can provide high fuel flexibility and extremely low emissions. However, the steam engine can run at a lower combustion temperature. In addition, the steam engine can be combined with a steam buffer, which is a high-temperature sensible heat storage. The steam buffer offers short-term energy storage, which can be useful, for example, if combined with a wind turbine in a wind-steam system for a stand-alone power supply.

2.7. Wave power

The power of waves is apparent to all who have been at the sea when the wind blows. Even when the wind calms down the waves continue to beat the shore. The waves are almost always in motion. But the energy content, of course, varies much, but not as much as in the wind. Wave power is to some extent smoothed wind energy. The wind affects a wave over a long distance and that gives an accumulation. The average power of the waves of the Baltic Sea is 2–6 kW per meter of wave front and in Kattegat it is 3–6 kW per meter. In the North Sea off Denmark, medium impact is up to 15 kW per meter and outside Norway, it can be as much as 50 kW per meter. The highest impact out on the great oceans is close to 100 kW per meter. For the Atlantic Ocean an estimated average is 60 kW per meter, but it may be 1700 kW per meter in stormy weather and 1 kW per meter at low wind (Fig. 2.61).

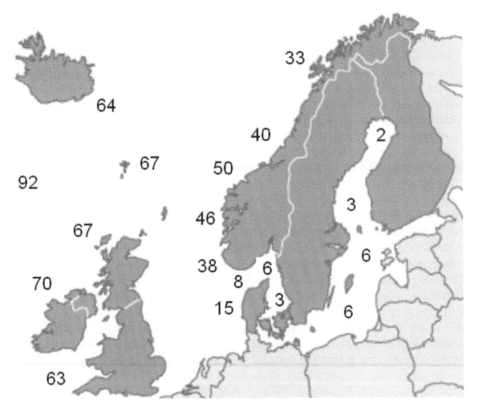

Figure 2.61 The average power of sea waves (kW/m) in northern Europe varies between 2 and 92 kilowatts per meter wave front.

2.7.1. *Properties of waves*

It's the big waves that can provide the greatest impact. The energy of the waves depends on their height, velocity and wavelength. The size of the waves depends on the wind speed, the distance that the wind can act on the wave and the depth and bottom topography (focusing or spreading wave energy). For a given wind speed, there is an upper limit to how big the waves can become, regardless of distance and time of impact. Wave power is usually measured in kW per meters wave front.

The northern and southern temperate zones of the globe have the best areas for wave energy. In the north, the predominant west winds are strongest in wintertime.

The water in waves moves in a circle – it rolls up. The movements of the waves are strongest at the surface and decrease exponentially with depth, but wave energy is also available as pressure waves at greater depths. Most of the energy is relatively close to the surface. Up to 95% are at a depth of a quarter of the wavelength.

When the waves come into shallower water, the movements slow down and energy is lost. The waves also pivot towards the shallower area. A wave with the energy of 50 kW per meter of wave front in deep water has maybe 20 kW or less when it reaches the beach. Wave power on beaches should lie in areas with large depths all the way into the beach.

The energy the waves carry is both potential and kinetic energy, often equally divided between the energy forms. The circulating movements in the water have kinetic energy, while water mass between crest and trough contain potential energy. The potential energy also creates pressure differences in the water below the surface, which can also be caught by certain installations.

Different wave power plants assimilate the different proportions of energy forms. A buoy that moves up and down captures kinetic energy. A wash-over facility where the water is collected in a higher-elevation reservoir exploits both potential and kinetic energy. The plant type "Salter's duck" (official name the Edinburgh duck), where a float rotates about an axis, is considered the most efficient of all structures, when it comes to capturing both potential and kinetic energy. It can capture 90% of the energy in waves. The sea becomes nearly stationary behind the plant. Development and commercialization of the facility has still not been successful.

2.7.2. Challenges

Many people have been fascinated by the wave force and sought to find ways to capture the energy. After the first oil crises in 1973, numerous wave energy programs were launched in many countries. But the task proved to be difficult, and most power plants started by the pioneers are defunct. Once again, though, many promising pilot plants are on their way to becoming commercial plants in full scale.

One difficulty is the environment. The facilities should be able to withstand the damp and salty atmosphere at sea. Typical problem areas are welds that crack and moorings that tear. In addition, the facilities must be dimensioned to withstand the worst winter storms. Because of this the plants have often been built so oversized that the economy has been devastated. The sea has also the tide that complicates structures further, especially for shore-located facilities.

A major challenge is capturing the energy, to efficiently convert wave movements into electrical energy. The waves vary continuously in height, length and speed. The movements are converted to generate electricity, which is easiest to do in rotating generators with speeds significantly higher than the rhythm of the waves. Plants' contact with the outside world is also a challenge. If a facility or any part of it is floating, it means using anchoring wires and electrical cables, which are constantly in motion with varying loads. This imparts a high risk of fatigue failure in the material.

2.7.3. Technical solutions

Many concepts exist for how to convert wave movements into useful energy. Wave power is characterized by the way power plants dampen the waves, how the energy is converted and where they are located. The following are some methods for harnessing wave power:

- Oscillating water column (OWC): The OWC is a hollow structure with an opening below the water surface. The water level changes with waves and water movement affects a volume of air which flows out and in. The stream of air is captured by an air turbine.
- Overtopping device: The water is led into a narrowed channel and may flow into a higher-lying reservoir. The energy is taken up with a hydro turbine with low-altitude demands.

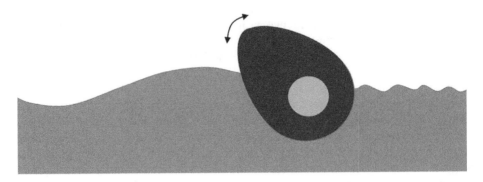

Figure 2.62 "Salter's duck" is a device that captures wave energy by moving up and down around a central axis. The nodding motion drives hydraulic pistons that convert the movement to rotation and produce electrical energy in a generator. The duck dampens the wave motion almost entirely, which is a sign of great energy uptake [51].

Source: G. Sidén.

- Point absorber buoy: Buoys drive linear generators and electricity is obtained directly. Lifting bodies can also drive hydraulic pumps that provide rotation in a hydraulic motor/ generator.
- Surface attenuator: The waves can bend a structure that pumps hydraulic oil, which can drive a hydraulic motor and generator.
- Rotating/tilting wave converter: The movement drives a pump and the flow is converted to rotation that drives an electric generator. One type generates a rotary motion directly which drives a generator (Fig. 2.62).
- Submerged pressure differential transducers: This is a relatively new technique that uses flexible (usually reinforced rubber) membranes to extract wave energy. These transducers use the difference in pressure at different locations below a wave to provide a pressure differential within a closed liquid system. The pressure differences are used to generate a flow that drives a turbine and an electric generator.

Facilities may be placed in the open sea or on shore. Sea-placed power plants have larger effects, but are obviously more vulnerable to storms. Shore-situated facilities should be located on steep banks, because long, shallow bottoms dampen the waves.

2.7.4. Wavestar – a Danish wave power concept

The Danish wave power project, Wavestar, is considered to be closest to a commercial launch. Wavestar's experimental facility in Hanstholm has produced electrical energy, but it is not yet on commercial basis.

The Wavestar power plant (Fig. 2.63) draws energy with floats, which rise and fall with the waves. The floats are connected by link arms to the platform, standing on legs fixed to the seabed. The arms move a hydraulic pump which provides the rotation of a hydraulic motor connected to an electric generator.

Figure 2.63 Wave Star is the wave concept in Denmark which is considered to be closest to commercial launch. The photo is from the pilot plant at Hanstholm [52].

Source: Wave Star A/S.

The test facility has only two floats; commercial plants will have many floats that will provide smooth, continuous energy. The first commercial Wavestar facility will have 20 floats, each 5 meters in diameter and 10 meters long, with link arms that transfer the wave forces. The nominal power is 600 kW, but plants in the megawatt size are included in development plans. At sea, the Wavestar power plants can be located together with off-shore wind power.

2.7.5. Wave Dragon

Wave Dragon was originally a Danish company but is now moving the center of their global operations to South Wales.

The test plant consists of a large floating barge, where electricity is produced by water floating up in a basin. Wave Dragon meets the ocean waves with long tentacles, which concentrate the waves up to a ramp that is relatively short and steep to minimize losses during the slowdown. The waves are collected in a water basin. The water flows back to sea level by conventional hydropower turbines, and electricity is generated (Fig. 2.64).

The purpose of the design is to make it as simple and reliable as possible. The only moving parts are the hydro turbines. For electricity generation a proven technology was chosen, Kaplan turbines, which work well even with low level differences. For the whole structure, standard components were selected that are well proven and cheap, providing inexpensive spare parts and maintenance costs. The floating height of the plant is variable, with tanks in which air can be pumped so that height is optimal for the current wave height.

The plant was also designed to withstand extreme waves. Big waves can simply rinse over the whole structure. Extreme winds can be handled by lowering the plant, so that it floats right above sea level.

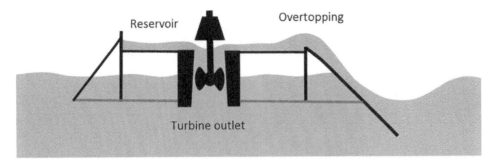

Figure 2.64 In the Wave Dragon the water is led towards a ramp where it is collected in a basin at a higher level than the sea surface [53].

Source: G. Sidén.

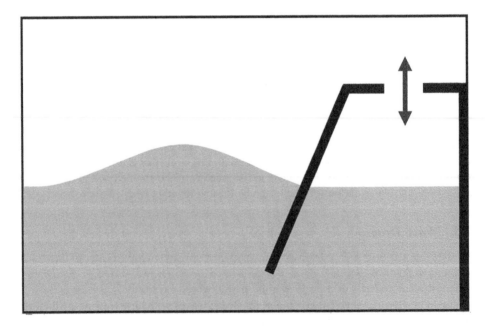

Figure 2.65 The OWC technology (oscillating water column) changes the water level in the air chamber with the waves. A stream of air flows out of and into the chamber. It is utilized in an air turbine for the generation of electricity [54].

Source: G. Sidén.

2.7.6. Limpet

Limpet (Land Installed Marine Powered Energy Transformer) is a facility of the OWC type, to be placed on the beach. It is constructed of an air chamber of concrete with an opening below sea level (Fig. 2.65). The waves produce air flow in and out of the chamber. This is used to drive an air turbine. A Wells turbine is used to drive the generator. The Wells turbine rotates in the same direction regardless of the air flow direction.

The first facility on the island of Islay, off Scotland's west coast, Limpet 500, with power of 500 kW, was installed in 2000 and has supplied energy to the national grid. The plant was later decommissioned, and in 2018 all installations except the concrete wave chamber had been removed. Based on this design the Mutriku Breakwater Wave Plant with 16 turbines has been built in the Bay of Biscay in Spain.

2.7.7. Buoy with linear generator

A wave power concept, buoy with a linear generator, has been developed in Sweden. The company Seabased has installed a test facility 2 kilometers west of the peninsula Islandsberg in the municipality of Lysekil. The Swedish Energy Agency and the power company Vattenfall have invested in the project.

The basis of the concept is a floating buoy and a linear generator. The buoy follows the movement of the waves and transmits the energy through a steel wire to generators on the seabed that then convert the energy to electricity.

The generators are placed in groups with a distance of 25–50 meters between each at a depth of 20–100 meters. Cables on the bottom connect the units to a collection center. The generated alternating currents are rectified and brought into shore with DC cables. The last step is inverters for feeding into the grid.

The first buoy in full scale (3 meters in diameter) was laid out in March 2005 along with a foundation of 40 tons. The generator's power is 10 kW. The whole concept is based on the newly developed linear generator. They are well protected on the bottom of the sea and run by the buoy with a steel wire. The system with the buoy, wire and generator can be seen as the simplest possible facility for wave energy. The number of units may vary from a few units to thousands connected to large ocean energy parks. The structure of the modules provides great flexibility. The system is intended for the relatively poor energy wave climate surrounding Sweden, but planned to still be able to produce electricity at a competitive cost (Fig. 2.66).

Seabased intended to build a large-scale wave power plant based on this concept in the same district. The plans have now been changed. Instead Seabased plans to establish production facilities in Brevik, Norway, which has a much better harbor. The Brevik facility will build and test wave energy converters (WECs) that will be part of their upcoming projects, including a 100-MW installation in Ghana in West Africa.

Another facility with a buoy is developed by the Swedish company Corpowers Ocean AB. The unit is connected to the seabed using a taut wire. The buoy movements drive gear wheels in the gearbox and rotation is produced which then drives electric generators.

The buoys are 8 meters in diameter. With a rated power of 250kW per unit in a typical Atlantic coast climate, this enables 10–250 MW wave farms with 40–1000 units. The concept allows mass production to drive down cost per unit and a maintenance scheme based on replacement of entire units at sea. This avoids service activities in the harsh off-shore environment, and it offers improvements in the operation time and reduction of cost for service and maintenance.

2.7.8. WaveRoller

The Finnish company AW-Energy Oy, partly owned by the energy company Fortum, has developed another wave concept – WaveRoller (Fig. 2.67). The company estimates that there are plenty of coastal bottoms with a depth of around 15 meters that are suitable for the aggregate type.

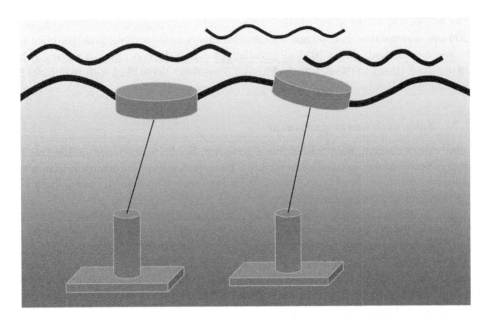

Figure 2.66 Buoy with direct energy conversion on the sea bottom with a linear generator – a Swedish concept for wave power [55].

Source: G. Sidén.

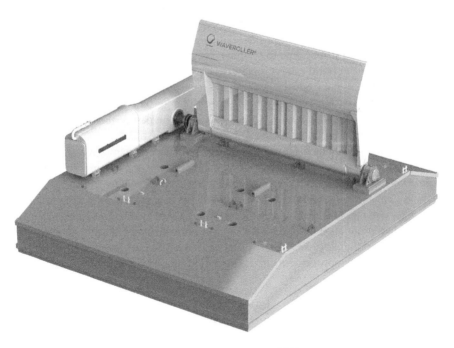

Figure 2.67 WaveRoller, a Finnish wave power concept [56].

Source: AV Energy OY.

The machine operates in near-shore areas, approximately 0.3–2 km out from the shore. The units are anchored to the seabed. Depending on tidal conditions it is mostly or fully submerged under the sea surface. A WaveRoller unit has a rated power of 350 –1000 kW and a capacity factor of 25–50%, depending on wave conditions at the site. The technology can be used as single units or in farms.

The back-and-forth movement of the seawater puts the WaveRoller panel into motion. To maximize the energy that the panel can absorb from the waves, the device is installed underwater at depths of approximately 8–20 meters, where the wave surge is most powerful.

As the WaveRoller panel moves and absorbs the energy from ocean waves, hydraulic piston pumps at the panel pump hydraulic fluids inside a closed hydraulic circuit. All the elements of the hydraulic circuit are enclosed inside a hermetic structure in the device and are not exposed to the marine environment. Consequently, there is no risk of leakage into the ocean. The high-pressure fluids are fed into a power storage and smoothing system, which is connected to a hydraulic motor that drives an electricity generator. The electrical output from this renewable wave energy power plant is then connected to the electric grid via a subsea cable.

The WaveRoller technology provides some unique benefits:

- The technology is installed and operates near shore, and is easy to access and is protected from extreme conditions, so infrastructure costs are low.
- It has a bottom fixed panel with a highly efficient power capture and can operate in low-, mid- and high-sea states with no stops and with only one moving part.
- It converts the wave movement to electricity using an onboard power storage and hydraulic system that gives smooth and grid-compatible power output.

2.7.9. Environmental impact

Wave power is basically a renewable energy source that contributes to reduced emissions from fossil energy. Moreover, the power plants are often located far off-shore, and at great depths, so they do not compete much with other uses of the sea.

When constructing the plant foundation, attachment points for securing wires and cable digging will stir up sediment, which may impact sea flora and fauna. However, it is a temporary problem and foundations can become artificial reefs, which could be positive for biological activity.

The wiring to its connection point with the electric grid can also affect the sea bottom and land.

Floating wave power devices are often low in the water and give no or insignificant visual impact.

Of course, the facilities should be located with respect to shipping interests, fisheries and recreation. But the sea is large, and it should be possible to find locations where interference is avoided.

The operation of the turbines can cause vibration or sound in the water. This impact should, of course, be examined. Research in this area is made today, even with regard to off-shore wind power. Noise from wave power should not be a problem on land. Plant noise levels are low, and become masked by the sound of waves.

Many wave power plants use hydraulic energy conversion. This involves the risk of oil spills. The oil quantities are normally limited.

Wave energy can also be used for desalination plants. They are located at the right place, close to beach areas where there may be a shortage of fresh water. The possibility of contributing to an emission-free supply of fresh water can, of course, be seen as an additional environmental advantage. Wave power, localized with the environmental concerns, should contribute to a better environment.

2.8. Tidal power

Tides cause the sea level to rise and fall periodically. The phenomenon is caused by the gravitational forces from the moon and to some extent from the sun and the earth's rotation (Fig. 2.68). The moon is held in its orbit by the attraction to the earth. But it also results in a physical impact on the Earth, with the ocean's water masses drawn against the side of the earth that is closest to the moon. On the side of the globe that is farthest from the moon its gravitational pull is lower, so even there the level rises. The Earth's rotation leads to that stream of water in the oceans flowing to the side where the moon is and to the opposite side of the earth. When the sun and moon are positioned in the same direction from the earth, an extra strong tide occurs, the spring tide.

The source of the currents is the earth's rotation, so the source of tidal energy is the stored rotational energy of the earth.

The tide leads to the rotation of the earth being slowed down; days will become longer. However, it is an extremely slow process and the earth's rotational energy is huge; we do not need to be worried about this.

An advantage of tidal energy is that – unlike many other renewable energy sources – it is largely independent of the weather. The tide occurs with great regularity, and production can be anticipated. When tidal power plants are built with systems of pools, the energy can also be stored for short periods, and production can start when energy needs are greatest.

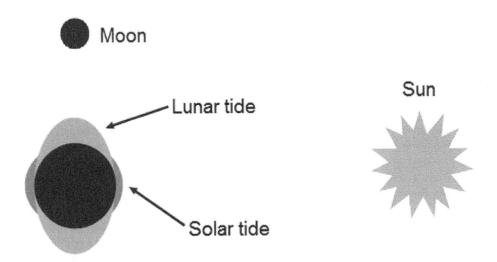

Figure 2.68 The tide is caused by an interaction between lunar and solar gravitational forces, the water masses of the ocean and centrifugal force caused by the earth's rotation.

Table 2.12 Tidal power plants in operation or in the planning phase [57]

Power plant	Power (MW)	Country	Operation year
In operation			
Sihwa Lake	254	South Korea	2011
River Rance estuary	240	France	1966
Annapolis Royal	20	Canada	1984
Pentland Firth tidal power plant*	6	Scotland	2018
Jangxia Creek, East China Sea	3.2	China	1980
Kislaya Guba	1.7	Russia	1968
Uldolmok	1.5	South Korea	2009
Eastern Scheldt	1.25	The Netherlands	2015
Bluemull Sound	0.6	Scotland	2017
In design and planning			
Incheon Tidal Power	1320	South Korea	
Garorim Bay	520	South Korea	
Severn Barrage	8640	England/Wales Russia	
Tugurskaya	3640	Russia	
Mezenskaya	≤12,000	Russia	
Penzhinskaya	87,100	Russia	
Swansea Bay Tidal Lagoon	320	Wales	
Skerries Tidal Stream Array	10.5	Wales	
Gulf of Kutch	50	India	
Alderney Tidal Plant	300	Alderney/Guernsey	

*Plans are made for 398 MW.

Along many coasts the difference in the level between high and low tide can be great.

Tides arise in daily or half-day periods. Half-hour periods, lasting for 12 hours and 25 minutes, are most common. Earth's total tidal power is estimated at around 3000 GW. Up to 400 GW are considered to be technically possible to extract. Today, 522 MW are extracted (Table 2.12). That will be tripled when Incheon in South Korea is completed.

2.8.1. Methods for energy collection

Two methods have been used to capture tidal energy. One is the barrier or lagoon plant that uses the potential energy that the tide can bring. By a barrier with controllable gates across a bay, a river estuary or a lagoon, a level difference between the water surfaces can be obtained. When the differences in the water levels are large enough – at least 3 meters – energy can be captured with a hydro turbine specially made for low level differences. Sometimes, a bi-directional turbine is used, that can capture the flow in both directions of the tides. A big improvement with longer operating times can be achieved by creating two or more pools.

The method requires substantial bays, so that large basins can be created. Conditions are suitable for about 5% of the world's coastlines. The sites are widely spread, and most of them are far from the locations where energy is needed. The first power plant was built in the 1960s at the mouth of the bay to the French river Rance. A power plant planned for the Severn estuary at England's west coast has been considered. However, the method carries a

Figure 2.69 Park with water turbines "Rotech Tidal Turbine – RTT" that can capture the energy in tidal currents at a great depth.

Source: Lunar Energy Ltd.

significant environmental impact, with major changes occurring to conditions for animals, plants and aquaculture. The disadvantages have led to postponing the Severn project.

The second method, ocean current power, captures kinetic energy directly from streaming water. The turbine can be designed as a windturbine-like construction under water (Fig. 2.69). Water is 800 times denser than air, so the power flow gives much larger power in relation to the swept area of the turbine in the water. The visual impact is small, the plant is not visible from the surface and there are no ecological changes that occur in dammed areas.

A third method that is suggested is dynamic tidal power (DTP). It is an untried but promising technology that would exploit an interaction between potential and kinetic energies in tidal flows. It proposes that very long dams (e.g. 30–50 km in length) be built from coasts straight out into the sea, without enclosing an area.

Along many coasts of the world, the main tidal movement runs parallel to the coastline: the entire mass of the ocean water accelerates in one direction, and later in the day back the other way. A DTP dam is long enough to exert an influence on the horizontal tidal movement, which generates a water level differential (head) over both sides of the dam. The head can be converted into power, using a long series of conventional low-head turbines installed in the dam.

2.8.2. Barrier tidal station in Rance

Tidal-driven mills have been used in areas with high tide since ancient times. One example is at the estuary of the river Rance in France (Fig. 2.70). Here the level differences are in average 8 meters with a maximum of 13.5 meters. The idea of a power plant was launched already around 1920, but it was first in the 1960s that construction started.

The costs of the power plant were approximately € 94.5 million, and it was inaugurated in 1967 under President Charles de Gaulle.

The plant has 24 turbines with a total peak power of 240 MW. The annual production is about 500 GWh [58]. The dam construction at the estuary totals 750 meters, and it is topped with a road. The total dammed area is 22.5 square kilometers. The power plant turbines are

Figure 2.70 The tidal power plant on the river Rance was the world's first power plant that retrieves energy from tides. It is located at the estuary of the river Rance in Brittany in France.

Source: Wikimedia Commons, CC-BY 2.5 license (https://creativecommons.org/licenses/by/2.5/deed.en).

designed to both generate electricity and pump water. The turbines are also reversible; they can be used both to flow water into the basin and out from it. The system is designed so that electricity can be produced continuously regardless of the timing of the tides.

The investment cost of the plant was high, but now with that outlay fully paid, the cost of the electricity is low.

The dam has affected the river's ecosystem. Sand eel and plaice have disappeared, but sea bass and squid have returned to the river. The operator, the French power company, EDF, is trying to adjust the levels to minimize the ecological impact.

The tidal power plant is a significant tourist attraction and has thousands of visitors per year. A gate in the west end of the dam allows the passage of ships between the English Channel and the Rance. By means of an openable bridge, large ships can pass through the barrier.

2.8.3. Tidal lagoon power

Around Britain's coasts are some of the best tidal resources in the world – water levels vary by 10–11 meters. The proposed project "Tidal Lagoon Power" was launched to collect energy from the tidal currents [59].

The proposal is to build six large tidal power plants along the British coast in Wales and England. The uniqueness of the project is the long ramparts that create large artificial lagoons (Fig. 2.71). If all six projects are built, they will provide 8% of Britain's entire electricity needs. In the long-term plans, there is tidal power with a total capacity of 7300 MW.

With the help of lagoons framed by kilometers-long ramparts that are up to 20 meters high, the immense power of the tides will be converted to electricity.

A proposed project in Swansea Bay outside Wales could generate almost 600 GWh of electricity per year. The lagoon would cover an area of 11.5 square kilometers trapped by a 9.5-km-long sea wall. At high tide the wall should reach 3.5 meters higher than the water surface.

Between the two outer ends of the walls, the tidal power plants would be built. They would consist of several-hundred-meters-long concrete walls with gates that provide openings for turbines.

When the tide begins to rise, the gates in the wall would be closed, creating a level difference into the lagoon. The gates would be opened when the water level becomes high enough, and water would stream out through the turbines, producing electricity while the lagoon is filled. When the tide turns, the gates would be closed until the level difference from the lagoon to the sea is high enough to produce an outflow.

The turbines capture the energy from the two incoming and two outgoing water flows per day, and are expected to produce an average of 14 hours a day. For power companies tidal energy has a major advantage in that production, unlike from solar and wind, is predictable in time.

The project was listed as part of the UK government's National Infrastructure Plan for 2014 and was granted planning permission in 2015. In early June 2018, the Welsh government offered to invest £ 200 million to improve the project's difficult financial situation. However, in June 2018, the Business, Energy and Industrial Strategy (BEIS) Department announced that a contract for an electricity trade agreement needed to fund the € 1.3 billion proposal was rejected by the government. The main reason was that lagoon tidal power would cost more for the average electricity consumer than what a mixture of wind and nuclear power would cost.

Other possibilities for carrying the project further were investigated. In February 2019, it was reported that the Swansea plan was revived – without the need for government funding.

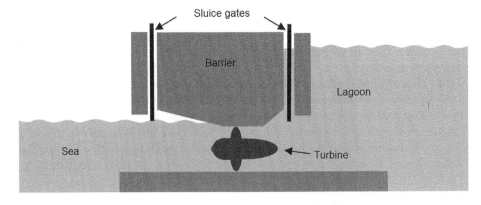

Figure 2.71 Principle design for a hydroelectric plant where the water is trapped in a bay or lagoon. The turbine captures the energy of water streams both in and out of the lagoon.

Source: G. Sidén.

The Swansea-based company, Tidal Power PLC, announced that several major companies were interested in buying the low-carbon electricity generated by the tide flowing through turbines in a concrete wall along Swansea bay.

The lagoon's supporters also believe that the project's prospects will improve by adding floating solar panels in the lagoon, which increases the amount of generated electricity. The supply of solar power should increase the annual energy production by more than one third, up from 572 GWh to about 770 GWh.

The plan is to secure enough funding to enable a final investment decision at the beginning of 2020, and that shortly thereafter construction should be able to begin.

2.8.4. Tidal power off Scotland

Between the Scottish mainland and the northern tip of the island of Stroma the tidal currents are powerful. A consortium led by the company MeyGen Ltd. plans here to build a large park with turbines for flowing water.

In early 2015, the first phase (1A) started. MeyGen had a development strategy, and four turbines were installed on gravity turbine support structures. The turbines each had the power of 1.5 MW, so a total of 6 MW were installed [60].

The turbines consist of a three-blade rotor mounted on a horizontal axis (Fig. 2.72). The rotor diameter is 18 meters and is fully submerged to a distance of at least 8 meters to the

Figure 2.72 A flow tidal turbine developed by the company Atlantis Resources lifted into place. Up to 269 turbines can be lowered to the seabed in the Pentland Firth.

Source: Atlantis Resources, name changed to Simec Atlantis Energy.

water surface. With a 4.4-kV submarine cable the turbines are connected to a power conversion center. It contains converters, control equipment, transformers and switchgear to take care of the turbine output and forward it on to the grid.

In February 2019 we could read that "A tidal turbine array in the north of Scotland set a new world record for generating power and exporting it into the national grid" [61]. The four turbine set-up in the Pentland Firth has generated 12 GWh of electricity since it was switched on last April – enough to power almost 9000 homes.

The second phase (1B) of the MeyGen project will involve the deployment of an additional four 1.5-MW turbines installed on innovative foundations. One goal is to significantly reduce the levelized cost of energy (LCOE) from tidal generation.

In the next phase (1C), 49 turbines will be built at MeyGen at an estimated cost of £ 420 million. The installations start in 2019. The park then will have 57 turbines with a total power of 86 MW [60].

In phase 2 and 3 the rest of the project will be developed. The goal is to install turbines with a total power of 398 MW.

Tidal power can significantly contribute to Scotland's target of generating 100% of its electricity with renewable energy sources by 2020. New research at the universities of Oxford and Edinburgh has identified suitable sites for tidal turbines off Scotland's north coast. A total of 1.9 GW could become installed. Production from these would be about 16 TWh per year. As a comparison, the electricity consumed in Scotland was 36 TWh in 2017.

2.8.5. Deep Green – Swedish tidal power

The company Minesto in Gothenburg, Sweden, is developing a tidal power unit for large depths out in the sea. Their technical concept, Deep Green, can be used at depths between 60–120 meters. Deep Green looks like a flying dragon and operates with large sweeping movements in the water (Fig. 2.73). A large wing carries the turbine generator. Deep Green is attached to the seabed by a wire. The sweeping movements increase the relative speed up to 10 times, enabling Deep Green to generate electricity even in slowly moving tidal and ocean currents.

Deep Green is planned to be manufactured in four different sizes with wingspans of 8–14 meter. The model DG500, 0.5 MW has a 12-meter wingspan. The length of the wires is 80–120 meters and the total weight is less than competing technologies.

The first test of Deep Green in an authentic ocean environment was completed at the end of 2011. In 2013, a prototype deployed in Northern Ireland has since been undergoing extensive testing in the tidal streams of Strangford Lough. The test units have a 3-meter-long wing and a generator that provides 3 kW. This will be achieved when the water speed is modest, 1 meter per second. In Strangford Lough, Minesto can use an area of 1.4 hectares. It gives Deep Green, which is anchored at 28 meters deep to a 40-ton concrete block, a sweep space of 120 meters in all directions.

Minesto's research and product development plan is to continue prototype testing in Northern Ireland, and in parallel build and deliver the first commercial-scale 0.5-MW Deep Green device in Holyhead Deep, Wales. The first installation will be followed by a gradual expansion to a 10-MW array. In February 2017, Minesto announced plans to expand the future capacity at Holyhead Deep to 80 MW [62].

Figure 2.73 Minesto's Deep Green captures the energy of the movement of water at great depths with sweeping movements. The side movements enhance water flow through the turbine.

Source: Minesto.

In autumn 2014, the marine energy technology companies Minesto and Atlantis Resources Ltd together were awarded € 750,000 from the EU Eurostars Programme. Funding is granted for a project which aims to reduce the costs of tidal energy.

It is a unique collaboration between two different marine energy developers in the global arena. The amount will be used to reduce the cost of tidal power by developing cost-effective components with high reliability of tidal turbines, for example turbine blades and carrying wings from composite materials.

2.8.6. The economy of tidal power

It has been considered difficult to harvest energy from the tides at competitive prices. Tidal power often requires large investments in facilities.

Dams for tidal power may sometimes be used for other applications such as bridges and ports. Thereby, the costs can be allocated. Parts of plants in the form of dams can also, when they are constructed, have a very long life, perhaps several hundred years, so the customary depreciation periods might not be used in cost calculations. For the power station at Rance the cost of the produced electricity is today, because the plant is written off, lower than costs of electricity from the French nuclear power plants. Much of the technology is also carefully tested, often in hydropower, so it should be possible to achieve long operating times and low operation and maintenance costs. The new concepts of flowing power stations that now have been launched may also give lower costs. Tidal power with competitive costs ought to be possible to reach.

2.9. Internet websites

General:

IRENA Global Atlas for Renewable Energy, www.irena.org/globalatlas/
RE Explorer, www.re-explorer.org
Natural Resources Canada (see e.g. Our Natural Resources, Energy Sources & Distribution), www.nrcan.gc.ca

Solar power:

PVGIS, re.jrc.ec.europa.eu/pvg_tools/en/tools.html
PVWatts Calculator, pvwatts.nrel.gov
PVOutput (sharing and comparing PV output data), pvoutput.org
Bengts nya villablogg (in Swedish), www.bengtsvillablogg.info

Wind power:

World Wind Energy Association (WWEA), www.wwindea.org, small-wind.org
Global Wind Energy Council (GWEC), www.gwec.net
WindEurope, windeurope.org
Danish Wind Industry Association ("Knowledge" and "Windpower wiki" contain a lot of information and calculators), en.windpower.org/wind-in-denmark
Windpower Monthly News Magazine, www.windpower-monthly.com
Hugh Piggott's blog (books and courses for home-builder), scoraigwind.co.uk
Wind Empowerment, windempowerment.org
Wind energy school by Bergey Windpower Co, bergey.com/wind-school
The technical certification scheme for construction, manufacturing, erection, operation and maintenance of wind turbines in Denmark (for example, set the filter on the Type column to starts with SWT), www.vindmoellegodkendelse.dk/egv-web-en-se
Danish Energy Agency, Wind Power (data about operating wind turbines in Denmark, also small), ens.dk/en/our-responsibilities/wind-power
myWindTurbine.com, Small turbines - Easy analysis, www.mywindturbine.com
Nordic Folkecenter (e.g. publishes Catalogue of Small Wind Turbines), www.folkecenter.net

Hydropower:

IEA Hydropower (under Publications are e.g. Miscellaneous Small Hydro Resources), www.ieahydro.org

HydroHelp, hydrohelp.ca

2.10. References

[1] PVGIS – Photovoltaic Geographical Information System, JRC, European Commission, 2019. [Online]. Available from: http://re.jrc.ec.europa.eu/pvgis.html [Accessed 2019-05-28].

[2] Yingli Solar, "PV-modules, YGE 60 cells", 2019. [Online]. Available from: http://www.yinglisolar.com/en/products/15 [Accessed 2019-05-28].

[3] EIA – US Energy Information Administration, "Today in Energy", 2018. [Online]. Available from: https://www.eia.gov/todayinenergy/detail.php?id=34952 [Accessed 2019-05-28].

[4] pvXchange Trading GmbH, Bremen, Germany, 2019. [Online]. Available from: https://www.pvxchange.com/de/aktuelles/preisindex [Accessed 2019-05-28].

[5] IRENA – International Renewable Energy Agency, *Solar PV in Africa: Costs and Markets*, 2016. [Online]. Available from: https://www.irena.org/-/media/Files/IRENA/Agency/Publication/2016/IRENA_Solar_PV_Costs_Africa_2016.pdf [Accessed 2019-05-28].

[6] Mälarenergi, "En av Sveriges största solparker". 2014. [Online]. Available from: https://www.malarenergi.se/om-malarenergi/vara-anlaggningar/solparken/ [Accessed 2019-05-28].

[7] Solar Energy International, *Solar Electric Handbook: Photovoltaic Fundamentals and Applications*, 2nd edition. [E-book] Boston, MA: Pearson Learning Solutions, 2013.

[8] H. Häberlin, *Photovoltaik: Strom aus Sonnenlicht für Verbundnetz und Inselanlagen*, 2nd edition. Fehraltorf: Electrosuisse Verlag, 2010.

[9] P. Dvorak, "Max Bögl Wind Puts Turbine on THE Tallest Tower, 178m", *Windpower Engineering & Development*, 2017-10-27. [Online]. Available from: https://www.windpowerengineering.com/mechanical/towers-construction/max-bogl-wind-puts-turbine-tallest-tower-178m-blade-tip-246-5m/ [Accessed 2019-03-25].

[10] Vindtek AB. [Online]. Available from: http://www.vindtek.se/hur.html [Accessed 2019-03-25].

[11] Windforce Airbuzz Holding AB. [Online]. Available from: http://www.windforce.se/vindkraft-windstar3000.php [Accessed 2019-02-25].

[12] World Wind Energy Association (WWEA), "Statistics", 2019. [Online]. Available from: http://wwindea.org/information-2/information/ [Accessed 2019-03-25].

[13] Vestas Wind Systems A/S, "V90–2.0 MW at a Glance". [Online]. Available from: https://www.vestas.com/en/products/2-mw-platform/v90-2_0_mw#!about [Accessed 2019-03-25].

[14] World Wind Energy Association (WWEA), "Small Wind World Report", Update 2017 (online summary). Available from: http://small-wind.org/wwea-released-latest-global-small-wind-statistics/ [Accessed 2019-03-25].

[15] S. Apelfröjd, *Grid Connection of Permanent Magnet Generator Based Renewable Energy Systems*, Uppsala University, 2016. [Online]. Available from: http://uu.diva-portal.org/smash/get/diva2:1033398/FULLTEXT01.pdf [Accessed 2019-03-25].

[16] P. Gipe, *Wind Power*. White River Junction, VT: Chelsea Green Publishing, 2004.

[17] P. Fraenkel, R. Barlow, F. Crick, A. Derrick, and V. Bokalders, *Windpumps: A Guide for Development Workers*, ITDG Publishing, 1993.

[18] A. Wyatt, *Wind Electric Pumping Systems: Sizing and Cost Estimation*, Research Triangle Institute, 1992.

[19] S. Ruin, Å. Larsson, E. Ingebrand, *Slutrapport för förstudie av autonom avsaltningsanläggning baserad på vindenergi*, report J103126002A by ÅF-Industri & System AB, Sweden, 2003 (for Ångpanneföreningens Forskningsstiftelse).

[20] SP Technical Research Institute of Sweden (now part of RISE), *Test Summary Report Giraffe 2.0 Hybrid Wind-Solar Power Station*, 2016-06-29. [Online]. Available from: http://small-wind.org/quality/labels/ [Accessed 2019-03-25].

[21] J. Earnest, T. Wizelius, *Wind Power Plants and Project Development*, 2nd edition. New Delhi: PHI Learning, 2015.

[22] Energy Facts Norway, "Electricity production", 2018. [Online]. Available from: https://energifaktanorge.no/en/norsk-energiforsyning/kraftproduksjon/ [Accessed 2019-05-30].

[23] CEDREN – Centre for Environmental Design of Renewable Energy. 2014. [Online]. Available from: https://www.cedren.no/english/ [Accessed 2019-05-30].

[24] International Hydropower Association, "Hydropower key facts 2018". [Online]. Available from: https://www.hydropower.org/keyfacts2018 [Accessed 2019-05-30].

[25] *The World Small Hydropower Development Report 2016*, United Nations Industrial Development Organization, Vienna, and International Center on Small Hydro Power, Hangzhou. [Online]. Available from: http://www.smallhydroworld.org/ [Accessed 2019-05-30].

[26] Hugh Piggott's blog, "Powerspout hydro turbines", 2019. [Online]. Available from: http://scoraigwind.co.uk/powerspout-hydro-turbines/ [Accessed 2019-05-30].

[27] IRENA – International Renewable Energy Agency, "Renewable energy statistics", 2018. [Online]. Available from: https://www.irena.org/publications/2018/Jul/Renewable-Energy-Statistics-2018 [Accessed 2019-05-30].

[28] IEA – International Energy Agency, "Geothermal energy". [Online]. Available from: https://www.iea.org/topics/renewables/geothermal/ [Accessed 2019-05-30].

[29] REVE (Wind Energy and Electric Vehicle Magazine), "The status of global geothermal energy", 2013-09-27. [Online]. Available from: https://www.evwind.es/2013/09/27/the-status-of-global-geothermal-energy/36290 [Accessed 2019-05-30].

[30] World Energy Council, "World energy resources geothermal", 2016. [Online]. Available from: https://www.worldenergy.org/wp-content/uploads/2017/03/WEResources_Geothermal_2016.pdf

[31] EIA – US Energy Information Administration, "Geothermal explained". [Online]. Available from: https://www.eia.gov/energyexplained/index.php?page=geothermal_power_plants [Accessed 2019–05–30].

[32] T. Tartière, M. Astolfi, "A World overview of the Organic Rankine Cycle market", September 2017. [Online]. Available from: https://orc-world-map.org/docs/WorldOverview2017.pdf [Accessed 2019-05-30].

[33] Orkustofnun – National Energy Authority, Iceland, "Energy statistics in Iceland", 2017. [Online]. Available from: https://nea.is/the-national-energy-authority/energy-data/data-repository/

[34] Orkustofnun – National Energy Authority, Iceland, "Direct use of geothermal resources", 2013. [Online]. Available from: https://nea.is/geothermal/direct-utilization/ [Accessed 2019-05-30].

[35] IF Technology bv, Netherlands, "Off-grid renewable electricity production with MiniGeo", 2016. [Online]. Available from: https://www.iftechnology.nl/off-grid-electricity-production-with-mini-geo [Accessed 2019-05-30].

[36] Directive 2004/8/EC of the European Parliament and of the Council of 11 February 2004 on the promotion of cogeneration based on a useful heat demand in the internal energy market and amending Directive 92/42/EEC.

[37] Engineering Toolbox, "Fuels – higher and lower calorific values". [Online]. Available from: https://www.engineeringtoolbox.com/fuels-higher-calorific-values-d_169.html [Accessed 2019-03-10].

[38] G. Boyle (ed), *Renewable Energy: Power for a Sustainable Future*, Oxford University Press, 1996.

[39] InStove, "Cookstoves". [Online]. Available from: http://instove.org/cookstoves [Accessed 2018-12-29].

[40] J-E. Liss, *Brännved – energiinnehåll i några olika trädslag*, Dalarna University, 2005. [Online]. Available from: https://www.diva-portal.org/smash/get/diva2:522819/FULLTEXT01.pdf [Accessed 2019-02-13].

[41] Forest Research, "Typical calorific values of fuels". [Online]. Available from: https://www.forestresearch.gov.uk/tools-and-resources/biomass-energy-resources/reference-biomass/facts-figures/typical-calorific-values-of-fuels/ [Accessed 2019-02-13].

[42] Waste Management World, "Low quality landfill gas to energy projects in Poland select cleanergy stirling engine gasbox". [Online]. Available from: https://waste-management-world.com/a/low-quality-landfill-gas-to-energy-projects-in-poland-select-cleanergy-stirling-engine-gasbox 2014-12-04 [Accessed 2019-01-09].

[43] A. I. Firmansyah, I. Adilla, Y. Gunawan, "Bioethanol (E100) Utilisation In 5 kVA Generator", *Journal of Ketenagalistrikan dan Energi Terbarukan*, vol. 16, no. 1, June 2017. [Online]. Available from: http://ketjurnal.p3tkebt.esdm.go.id/ketjurnal/index.php/ket/article/view/152/144 [Accessed 2019-02-06].

[44] *Swedish Energy Agency, Report ER 2018:17*, revised edition, January 2019. [Online]. Available from: https://energimyndigheten.a-w2m.se/Home.mvc?ResourceId=5753 [Accessed 2019-02-06].

[45] Linköping University, "Lignin – a super green fuel for fuel cells", 2018-05-14. [Online]. Available from: https://liu.se/en/news-item/lignin-nytt-supergront-bransle-for-branslecell [Accessed 2019-01-29].

[46] L. L. J. Mahon, *Diesel Generator Handbook*. Oxford: Elsevier, 1992.

[47] Victron Energy, "Victron energy Marine Generator Test 2007", revised 2008-01-07.

[48] Energimyndigheten – Swedish Energy Agency, "Test av reservelverk från 2015". [Online]. Available from: http://www.energimyndigheten.se/tester/tester-a-o/reservelverk-2015/ [Accessed 2019-04-01].

[49] *Diesel Generators: Improving Efficiency and Emission Performance in India*, Shakti Sustainable Energy Foundation and ICF International, 2014.

[50] Ranotor Utvecklingsaktiebolag, "Modern high performance small-scale steam power". [Online]. Available from: http://www.ranotor.se [Accessed

[51] J. Taylor, "Edinburgh wave power group", 2009. [Online]. Available from: http://www.homepages.ed.ac.uk/v1ewaveg/ [Accessed 2019-05-30].

[52] Wavestar, "Unlimited clean energy with the wavestar machine". [Online]. Available from: http://wavestarenergy.com/ [Accessed 2019-05-30].

[53] Wave Dragon, "Prototype testing in Denmark", 2009. [Online]. Available from: http://www.wavedragon.co.uk/2009/03/15/prototype-testing-in-denmark/ [Accessed 2019-05-30].

[54] Inverness Courier, "Inverness firm hands over the world's first full life wave power plant", 2011. [Online]. Available from: https://www.inverness-courier.co.uk/News/Inverness-firm-alunches-first-tidal-wave-project-16112011.htm

[55] MarineEnergy.biz, "Seabased to set up production facility in Norway". [Online]. Available from: https://marineenergy.biz/2018/05/09/seabased-to-set-up-production-facility-in-norway/ [Accessed 2019-05-30].

[56] WaveRoller, "Company history". [Online]. Available from: http://aw-energy.com/about_us/#companyhistory [Accessed 2019-05-30].

[57] Processed from List of tidal power stations, Wikipedia, 2019. [Online]. Available from: https://en.wikipedia.org/wiki/List_of_tidal_power_stations [Accessed 2019-05-30].

[58] EDF, "Tidal Power: EDF a Precursor". [Online]. Available from: https://www.edf.fr/en/the-edf-group/industrial-provider/renewable-energies/marine-energy/tidal-power [Accessed 2019-05-30].

[59] Tidal Lagoon Power, "Harnessing the power of our tides". [Online]. Available from: http://www.tidallagoonpower.com/ [Accessed 2019-05-30].

[60] Simec Atlantis Energy, "MeyGen". [Online]. Available from: https://simecatlantis.com/projects/meygen/ [Accessed 2019-05-30].

[61] The Herald, "Meygen tidal turbine powers on to world record". [Online]. Available from: https://www.heraldscotland.com/news/17451080.meygen-tidal-turbine-powers-on-to-world-record/ [Accessed 2019-05-30].

[62] Minesto, "Project and technology development". [Online]. Available from: https://minesto.com/projects [Accessed 2019-05-30].

Chapter 3

Electricity storage

Storing energy, in particular in a form that enables a supply of electricity, can have many functions in small-scale renewable energy systems. A traditional use is in off-grid systems, where storage enables surplus generation to be stored to another period of time when there is no generation. In grid-connected systems as well as off-grid, storing energy can enable reduction in the peak power imported from the grid or a genset ("peak-shaving"). Grid-connected systems can use stored energy to be more self-sufficient with energy. On some markets, they can also benefit from participating in frequency regulation or storing renewable energy until a later time when imported energy from the grid is expensive. Improving power quality is another opportunity on the local level. Grid backup systems can include all of the foregoing. Furthermore, storing electricity is a key issue for transporting it in new ways.

The different types of electricity storage technologies are characterized by qualities such as energy storage capacity, useful energy storage capacity, cycle energy efficiency, energy to power ratio, self-discharge, energy and power density, maintenance requirements, lifetime, cost, safety and influence on the environment. The cycle efficiency is of importance in many ways. For example, 80% cycle efficiency means that 20% will be lost as heat, and in case of a large energy throughput there can be much heat (so cooling can be an issue). The losses can, of course, also affect the sizing of other systems, such as a PV array that charges the storage.

Some of the options for storing electricity are as follows:

Mechanical:

- Pumped hydro storage (PHS) is a field-proven technology for stationary applications, which can in suitable cases provide a low cost per kWh stored and long cycle life. However, it is normally not economical for the smallest systems, is dependent on a suitable geography, requires access to two reservoirs at different altitudes, can impact the environment and may require years of permitting and construction. Even though it can in some places provide storage for months, that is not always the case (see the case study "El Hierro: self-sufficient island with pumped hydropower" in Chapter 6).
- Compressed air energy storage (CAES) in underground caverns is also for stationary applications. Like PHS, it demands favorable sites and geological formations. It can be used together with gas turbines, powered also by natural gas, to extract the stored energy.
- Flywheels are mainly used to ensure short-duration power quality and uninterruptible power supply. A disadvantage is their very high self-discharge rate.

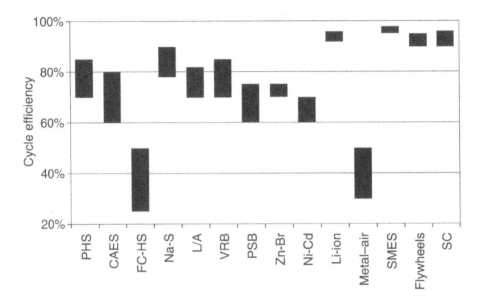

Figure 3.1 Cycle efficiency for some energy storage technologies, according to D.P. Zafira-kis [1]. PHS = Pumped hydro storage; CAES = Compressed air energy storage; FC-HS = Fuel cell and hydrogen storage; Na-S = Sodium-sulfur battery; L/A = Lead-acid battery; VRB = Vanadium redox battery; PSB = Polysulfide bromide battery; Zn-Br = Zinc-bromine battery; Ni-Cd = Nickel-cadmium battery; Li-ion = Lithium-ion battery; Metal-air = Metal-air battery; SMES = Superconducting magnetic energy storage; SC = Supercapacitor. Reprinted with permission.

Electrical:

- Supercapacitors (SCs), or ultracapacitors, are low-voltage components, presently available with capacitance up to thousands of farads, with high-cycle efficiency. Unlike batteries, their voltage drops significantly as they are discharged.
- Superconducting magnetic energy storage (SMES) stores energy in the magnetic field produced when direct current flows through a superconducting coil. For this to happen, the material of the coil must be cooled to a temperature below its superconducting critical temperature.

Chemical/electrochemical:

- Fuel cell and hydrogen storage (FC-HS) is a form of power-to-gas and gas-to-power energy storage, where electricity and water is used by an electrolyzer to produce hydrogen. The gas is stored and can then be converted back to electricity by a fuel cell (see the case study "Home with seasonal storage of hydrogen" in Chapter 6).
- Power-to-liquid usually means that electricity together with carbon dioxide and water is used to produce a liquid fuel, which can later be used, for example, for driving a vehicle or genset. One example is the German company Sunfire, which has produced a synthetic crude oil substitute, called "Blue Crude" [2]. Researchers at Chalmers University of

Technology in Gothenburg also see an opportunity in what they call electrofuels, like synthetic petrol [3].

• Flow batteries, such as vanadium redox battery (VRB), polysulfide bromide battery (PSB) and zinc-bromine (Zn-Br) battery, are sometimes also called regenerative fuel cells. Energy is stored in two liquid electrolyte solutions. The quantity of the electrolytes determines the energy capacity, which is independent from the rated power determined by the cell stack.

• Batteries can be primary batteries, designed to be used only once, or secondary batteries, designed to be recharged. The rest of this chapter focuses on rechargeable batteries, because that is the dominating technology in small-scale renewable energy systems today and is growing fast. Batteries are often found also in systems using another storage technology.

A common use of batteries in the smallest renewable energy systems is 12 V, 24 V and 48 V systems using lead-acid batteries. These are nominal voltages – actual voltage can be a little higher or lower. Nowadays also more advanced battery systems of much higher voltage, such as Tesla Powerwall, are available for the homeowner. A battery inverter can be used to convert the direct current (DC) of the battery to alternating current (AC), and it is often bi-directional.

An interesting potential market driver is multi-use of batteries. Today, some batteries are often used actively only during a minority of their useful lifetime. Reducing the time they sit idle and increasing their use rate can enhance the value they provide. It is possible to use batteries for more than one thing, for example using an electric car for transportation during part of the day and when it is parked for AC grid frequency regulation (see the case study "Municipal utility company using V2G and solar PV" in Chapter 6). Multi-use could also include the possibility of grid backup in case of a grid failure.

However, storing electricity the traditional way in, for example, lead-acid batteries has in some ways been the "Achilles' heel" (weakness despite overall strength) of small renewable energy systems. This is not only because of the direct costs associated. Often a worn-out small renewable energy system will be accompanied by a set of old batteries dumped in nature, where battery materials may be a source of pollution for coming generations. Heavy metals such as lead and cadmium will not disappear, if left unattended. This makes responsible recycling very important, also for old abandoned batteries. The experience of one of the authors is also that, for example, in developing countries, some batteries are of low quality or supposedly new batteries may have been used before. This is not surprising, because many counterfeit electrical components are sold on such markets. There can also be fraudulent battery ratings, which can even be ridiculous [4]. Some problems may also be caused by the lack of proper battery management systems.

A simple calculation of dividing the storage-system purchase price (say US$ 100 for a battery with 1 kWh gross capacity typically used for off-grid systems) with the energy delivered over its operating lifetime (in this example 1000 kWh delivered during 2000 cycles at 50% depth-of-discharge, which would be an excellent or even overly optimistic result for that price) shows that storing one kWh in such a battery costs at least US$ 0.1 (some associated costs are omitted in this simplified calculation) [5]. Such a cost is in many countries about the same cost as buying a kWh of electricity from the grid.

That may change in the future, not least because of the second-life of electric vehicle (EV) batteries. With the large number of EVs that can be expected in the world, there will be many used batteries no longer desired for their original purpose. Many of these batteries will probably be available for a low price and could still be useful for many years in other applications (typically stationary use where a reduced capacity per battery is of less importance). More environmentally friendly battery technologies have also been developed, for example based on salt water.

3.1. Battery basics

A battery consists of one or more electrochemical cells with external connections to provide electric power. A battery's capacity is the amount of electric charge it can deliver, usually measured in Ampere-hour (Ah), while remaining above a specified terminal voltage per cell. Capacity can also be expressed in terms of energy (Wh), that is, the discharge capacity in Ah multiplied by the nominal (average) voltage. For different battery types, the cell voltage can be from 1 to 4 V.

Battery cells can be grouped together in a battery block (also called battery module, mainly in vehicle applications). Usually, many cells or blocks are connected in series. For example, an ordinary 12-volt lead-acid car battery consists of six 2-volt cells in series. More than one such battery string can be connected in parallel, which will increase the total capacity in terms of current, amphour and energy (but not increase the total voltage). The entire battery is sometimes called a battery bank (or battery pack, mainly in vehicle applications).

In an ideal battery, all cells are identical and thus have the same cell voltage and capacity. In reality, that is not the case. As long as the differences are small among the cells, the battery will work fine. However, the differences tend to grow over time, unless the voltage difference is adjusted, which can be done by active or passive cell balancing. Without balancing, those cells with lower voltage than the average cell in the battery will receive less charge. Other cells with higher voltage than the average will receive more charge. Thus, we can end up with a battery that consists of some undercharged cells and other overcharged cells, while the total may falsely appear to be in good condition. Therefore, it is important to keep all cells of a battery as identical as possible – same type, age, state-of-charge (SOC), etc.

Just as a chain is not stronger than its weakest link, the battery will not be stronger than the weakest cell. Especially if a cell is damaged by under- or overcharging, it may break, which will cause that entire battery string to fail. To avoid a single point of failure, it may be tempting to use parallel strings. However, if one cell in such a battery bank has a lower voltage than average, the risk is that when charging is not taking place, all other strings will try to charge the string with that cell. This can lead to a battery bank that never stays fully charged after charging is ended.

The available capacity depends on several factors, including temperature and the rate at which charge is being delivered (current). As the current increases, the battery's available capacity decreases. The so-called C-rate is a measure of the rate at which a battery is being charged or discharged. A 1C discharge rate would deliver the battery's rated capacity in one hour. Using a battery rated 80 Ah as an example, a 1C discharge means a current of 80 A (which is a fast charge/discharge for that battery), 0.1C (C/10 or C10) means a current of 8 A, and so on. The battery capacity is usually specified at a certain discharge rate, for example 80 Ah at 0.05C. At a 1C discharge rate, the available capacity for the same battery would be less. To avoid the dimensional error of expressing a current in the unit Ah, the concept I_t was introduced by the international standard IEC 61434. I_t is equal to the capacity C divided by one hour. Thus, it is more correct to speak of, for example, "$2I_t$ rate" instead of "2C rate".

While the rated capacity is normally stated on the battery, the useful capacity is something else. Some types of batteries, like lead-acid, will wear out more quickly if they are

Table 3.1 Characteristics of commonly used rechargeable batteries, based on batteryuniversity.com. The figures are based on average ratings of commercial batteries at time of publication. Specialty batteries with above-average ratings are excluded [6]. Reprinted with permission.

Specifications	Lead-acid	NiCd	NiMH	Li-ion[1]		
				Cobalt	Manganese	Phosphate
Specific energy (Wh/kg)	30–50	45–80	60–120	150–250	100–150	90–120
Internal resistance	Very low	Very low	Low	Moderate	Low	Very low
Cycle life[2] (80% DoD)	200–300	1,000[3]	300–500[3]	500–1000	500–1000	1000–2000
Charge time[4]	8–16h	1–2h	2–4h	2–4h	1–2h	1–2h
Overcharge tolerance	High	Moderate	Low	Low no trickle charge		
Self-discharge/month (room temp)	5%	20%[5]	30%[5]	<5% Protection circuit consumes 3%/month		
Cell voltage (nominal)	2V	1.2V[6]	1.2V[6]	3.6V[7]	3.7V[7]	3.2–3.3V
Charge cutoff voltage (V/cell)	2.40 Float 2.25	Full charge detection by voltage signature		4.20 typical; some go to higher V		3.60
Discharge cutoff voltage (V/cell, 1C)	1.75V	1.00V		2.50–3.00V		2.50V
Peak load current Best result	5C[8] 0.2C	20C 1C	5C 0.5C	2C <1C	>30C <10C	>30C <10C
Charge temperature	−20 to 50 °C (−4 to 122 °F)	0 to 45 °C (32 to 113 °F)		0 to 45 °C[9] (32 to 113 °F)		
Discharge temperature	−20 to 50 °C (−4 to 122 °F)	−20 to 65 °C (−4 to 149 °F)		−20 to 60 °C (−4 to 140 °F)		

(Continued)

Table 3.1 (Continued)

Specifications	Lead-acid	NiCd	NiMH	Li-ion[1]		
				Cobalt	Manganese	Phosphate
Maintenance requirement	3–6 months[10] (topping chg.)	Full discharge every 90 days when in full use		Maintenance-free		
Safety requirements	Thermally stable	Thermally stable, fuse protection		Protection circuit mandatory[11]		
In use since	Late 1800s	1950	1990	1991	1996	1999
Toxicity	Very high	Very high	Low	Low		
Coulombic efficiency[12]	~90%	~70% slow charge ~90% fast charge		99%		
Cost	Low	Moderate		High[13]		

1 Combining cobalt, nickel, manganese and aluminum raises energy density up to 250 Wh/kg.
2 Cycle life is based on the depth of discharge (DoD). Shallow DoD prolongs cycle life.
3 Cycle life is based on battery receiving regular maintenance to prevent memory effects.
4 Ultra-fast charge batteries are made for a special purpose.
5 Self-discharge is highest immediately after charge. NiCd loses 10% in the first 24 hours, then declines to 10% every 30 days. High temperature and age increase self-discharge.
6 1.25V is traditional; 1.20V is more common.
7 Manufacturers may rate voltage higher because of low internal resistance (marketing).
8 Capable of high current pulses; needs time to recuperate.
9 Do not charge Li-ion below freezing.
10 Maintenance may be in the form of equalizing or topping charge* to prevent sulfuration.
11 Protection circuit cuts off below about 2.20V and above 4.30V on most Li-ion; different voltage settings apply for lithium-iron-phosphate.
12 Coulombic efficiency is higher with quicker charge (in part due to self-discharge error).
13 Li-ion may have lower cost-per-cycle than lead-acid.
* Topping charge is applied on a battery that is in service or storage to maintain full charge and to prevent sulfuration on lead-acid batteries.

often discharged to 100%, so much more shallow cycling is preferred. Often a safety margin against total discharge is also desired. Those are some reasons why the useful capacity can be less than rated.

Batteries normally obtain optimum service life at a temperature around 20 °C. Higher temperatures than that typically lead to shorter life. Lower temperatures typically lead to reduced available capacity and charge/discharge rates.

Good instruction documentation and customer support is not always available for batteries, but one good example is the Sonnenschein series of batteries, which has datasheets with extensive specifications, operating manual, handbook and knowledgeable support. An often overlooked parameter for batteries is the short-circuit current they can deliver. This is useful, for example, when selecting a fuse or MCB to be placed on the battery cable, where the interrupt capacity of that device should be higher than the possible short-circuit current. For example, a Sonnenschein SB6/200 (6V, 200Ah) battery has a short-circuit current of 2800 A, according to their support.

3.2. Lead-acid batteries

Lead-acid batteries (Fig. 3.2) are available in two main types:

- Flooded lead-acid (FLA) batteries have a liquid electrolyte, which is a mixture of sulfuric acid and water. Some of this water is released as hydrogen and oxygen during charging. Water may also be lost as a result of evaporation at high temperatures. Thus, the electrolyte level needs to be checked and pure water (usually distilled or deionized)

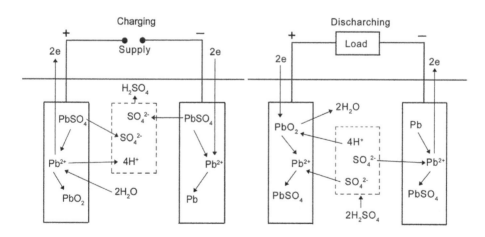

Figure 3.2 Working principle of a lead-acid battery cell. When a voltage is applied that is higher than the cell voltage, current will flow to the positive terminal (electrons to the negative) and the cell will be charged. Lead (IV) oxide is formed at the positive anode, pure lead is formed at the negative cathode and sulfuric acid is liberated into the electrolyte, causing the specific gravity to increase (left). During discharge, current will flow through the load, in the opposite direction. Lead sulfate is formed at both electrodes and sulfuric acid is removed from the electrolyte, causing the specific gravity to decrease (right). Based on copyleft illustration from ITACA [7].

needs to be topped up at certain intervals. There are also "recombination vent caps", reducing water loss and therefore the need for adding water.

- Valve-regulated lead-acid (VRLA) batteries, which are sometimes also less accurately called sealed lead-acid (SLA) batteries or maintenance-free batteries. Battery water cannot be topped up in them. However, they can still need maintenance, for example by cleaning, tightening of connections and functional testing. They can be either valve-regulated wet cell, gell cell or absorbent glass mat (AGM). VRLA batteries normally retain generated gases within the battery, so the gases can recombine within the battery itself (sometimes with the help of a catalyst). If the pressure exceeds a certain limit, then valves open to allow the excess gases to escape.

Hydrogen released during charging poses an explosion risk, unless the surrounding is properly vented and possible ignition sources are located at a minimum distance from the battery. Therefore, the battery fuses/MCBs are often located some distance from the battery. The positive and negative battery wires are preferably kept well separated until after the battery fuses/MCBs (to minimize the risk of short-circuit before the fuses/MCBs). Lead-acid batteries also suffer from toxic and harmful content. For some applications it is a drawback that the available discharge capacity is much reduced at high current and achieving full charge takes a long time (in particular the last few percent, and a shorter lifespan can be expected if full charge is not regularly achieved). However, some lead-acid batteries are designed not to be fully charged regularly, like Northstar PSOC. The freezing temperature of a lead-acid battery varies between approximately −70 °C for a fully charged battery to close to 0 °C for a fully discharged battery.

For lead-acid batteries, Peukert's law is often used as an approximation for change in capacity at different rates of discharge. It can be reformulated to give an estimation of the time t (in hours) that the battery will last, given a particular rate of discharge (current):

$$t = H\left(\frac{C}{IH}\right)^k$$

(3.1)

where: H is the rated discharge time (hours), C is the rated capacity (Ah) at that discharge rate, I is the actual discharge current (A) and k the Peukert constant (dimensionless).

The Peukert constant varies between 1 and about 1.6, where a value close to 1 shows that the battery performs well with high current.

The SOC of lead-acid batteries can be estimated, for example, by measuring the open-circuit voltage with a handheld multimeter, after the battery has been at rest for some time. See Figures 3.3 and 3.4 for one type of battery, which is (unlike ordinary starting batteries) designed for deep-cycle applications like off-grid power systems. However, when measuring block voltage rather than cell voltage, be aware of the uncertainty because the individual cells may differ within the block. Different types of lead-acid batteries will have different charge/discharge curves, but the principle is similar for all of them. In case of flooded cells, a hydrometer can be used to measure the specific gravity of the electrolyte for an approximation of SOC.

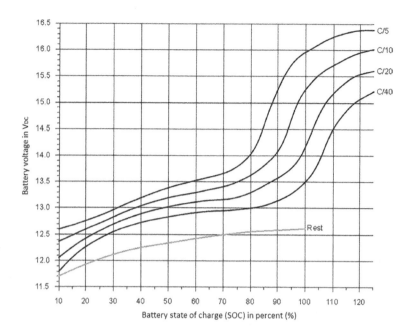

Figure 3.3 12-volt lead-acid battery state-of-charge (SOC) vs. voltage while battery is under *charge*, according to *Home Power*. Data presented here were generated from Trojan L-16W deep-cycle lead-acid batteries [8]. Reprinted with permission.

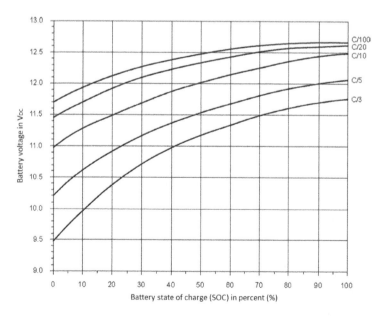

Figure 3.4 12-volt lead-acid battery state-of-charge (SOC) vs. voltage while battery is under *discharge*, according to *Home Power*. Data presented here was generated from Trojan L-16W deep cycle lead-acid batteries [8]. Reprinted with permission.

3.3. Nickel-based batteries

Nickel-cadmium (NiCd) batteries are known to tolerate deep discharge, provide nearly rated capacity also at very high discharge rates and be useful in extreme temperatures, in some cases −50 to +70 °C. Even though some types are resistant to overcharging, this may cause connected equipment to trip because of overvoltage. On the downside, they contain toxic cadmium, have a low energy density, can have a memory effect and low cycle efficiency. Both flooded NiCd cells that require watering and valve-regulated ("sealed") NiCd cells without a watering requirement exist. Especially the latter may require that overcharging be avoided.

Nickel-iron (NiFe) batteries have many similarities to NiCd, but don't have a memory effect and contain no toxic cadmium. However, they are less suitable for high charge/discharge levels, can have even lower energy density, lower cycle efficiency, higher self-discharge, more pronounced hydrogen gassing and require watering.

Nickel metal-hydride (NiMH) batteries (Fig. 3.5) are often considered as being relatively safe, and might therefore be treated as non-dangerous goods during transport. However, when there is much stored energy, there is always a possibility for fire or a battery melting. They must still be protected from short-circuiting and protected from movement that could lead to short-circuiting. Traditionally, NiMH batteries have been known for a high self-discharge. However, low self-discharge NiMH batteries are nowadays available. Modern NiMH batteries also have no memory effect to an extent that that will be noticed. There appears to be a myth that NiMH batteries need to be fully discharged before recharging them. However, some are very sensitive to overcharging and should therefore not be constantly "trickle charged." Recently, researchers have found that the life of NiMH batteries can be multiplied by re-conditioning them with oxygen [9].

3.4. Lithium-based batteries

The high specific energy (Wh/kg) of lithium-based batteries, compared with other types of batteries, makes them very attractive, especially in mobile or portable applications. Different types of rechargeable lithium-ion batteries (which should not be confused with non-rechargeable lithium iron disulfide batteries) have become widely used, for example in electric cars. They have very high cycle efficiency, good cycle durability and low self-discharge, and are suitable for high charge and discharge rates, except at low temperatures when especially the charge rate can be reduced or charging may not even be possible for some types. However, safety is a concern, for example during transportation. Lithium-ion battery fires can cause intense heat and smoke, and may even cause explosions. The toxic gases can be the most significant threat [10]. Two of the many chemistries of lithium-ion batteries are briefly describe next.

The lithium iron phosphate battery ($LiFePO_4$), also called LFP battery, is considered to be one of the safest types of lithium-ion battery. An example of an LFP battery block is shown in Figure 3.6 and the corresponding discharge curve in Figure 3.7. Such a flat discharge curve is ideal for many applications. For example, a battery inverter in an off-grid system is less likely to trip because of battery undervoltage when a large load is applied, even if such a battery is relatively small.

Another lithium-ion battery type considered to be one of the safest is the lithium-titanate battery (e.g. $Li_4Ti_5O_{12}$), usually abbreviated LTO battery. They are claimed to have an extreme cycle durability (in some cases more than 20,000 cycles), can be charged/discharged

Figure 3.5 Example of a modern NiMH battery from Nilar for residential energy storage. Total capacity of the three gray battery cabinets is 69 kWh. This battery system is used together with a large grid-connected PV system. The owner, Robert von Bahr, stands next to the power electronics cabinet from Ferroamp, which includes grid backup functionality.

Source: S. Ruin.

very quickly (in the order of $10I_t$, which means six minutes for a full charge) and have good low-temperature performance (down to e.g. –30 °C).

Unlike lead-acid batteries, some modern batteries like lithium-ion can achieve an increased cycle life by operating at partial SOC (e.g. 20–70 %) and not be kept fully charged too much. However, depending, for example, on the application, they may need to be fully charged not too seldom for the balancing. Thus, there are nowadays interesting opportunities for smart charging systems of a totally different kind than for lead-acid.

Many battery types, not least lithium-ion, can be severely damaged by over-discharge. Attention needs to be paid to electronics that might stay connected and slowly drain the battery also after a low-voltage disconnection of ordinary loads. (If over-discharge would still take place, it is the experience of one of the authors that LFP batteries have in some cases been rescued by very slowly charging each block separately and stopping charging as soon as full voltage is reached.)

Recycling of lithium-based batteries has so far not been widespread, but this situation is improving, which can be crucial for providing a fast-growing manufacturing industry with materials.

3.5. More than just a battery

Many batteries have some type of built-in protection and other functions, which can range from simple to sophisticated.

In addition to simple vents, there may be a pressure valve that will disable the cell permanently if the pressure is too high in the cell, for example because of over-charging. For example, in small cylindrical cells there can also be a PTC resistor (a temperature-sensitive resistor with a very low resistance at normal temperature, but at a certain temperature the resistance will increase dramatically) to protect against over-temperature and limit the current flow. A PTC will automatically reset itself when it cools down. Even some small cylindrical cells also contain a small printed circuit-board (PCB) with protective functions, for example against over-discharge, over-charge and over-current. The PCB current consumption will add to the self-discharge of the cell itself.

In many modern battery blocks, especially lithium-ion, a battery management system (BMS) or battery management unit (BMU) is integrated in the battery casing. An example of such an LFP battery and its discharge curve are shown in Figures 3.6 and 3.7. In addition to providing protective functions against misuse, this BMS constantly monitors and balances the battery cells. This battery can in many cases be used as a drop-in replacement for old VRLA/AGM batteries.

Figure 3.6 Example of an LFP battery with built-in BMS [11]. Reprinted with permission.

Because of the rather flat discharge curve for some types of batteries (as seen in Figure 3.7), it can be difficult to estimate SOC based on voltage at the battery terminals. Coulomb counting (also called amp-hour counting) is another way to estimate SOC, which means that the current flowing in and out of the battery is measured and integrated over time, preferably with losses considered. This can be combined with automatic calibration with the "chemistry" of the battery at some known point, such as zero setting of coulomb counting when the battery is fully charged. Coulomb counters may drift considerably, for example if a battery is seldom fully charged. Kalman filter is another way to calculate SOC [12]. In some cases a method to calculate and present the SOC is included in the BMS and in other cases it is external.

A BMS can also monitor and present the battery state-of-health (SOH), for which there is no consensus on how to determine. In any case, an SOH equal to 100% is assumed for a fresh/new battery. The SOH can be evaluated by comparing it to a threshold for the application in question.

Some systems may observe the charge time because a faded battery may charge "too quickly."

Many BMS systems also contain some type of communication between blocks, with a display, etc., for example via Bluetooth. Unfortunately these communications are often based on various proprietary protocols, which can make system integration more difficult.

A suitable BMS is required for many types of modern batteries and their applications, to achieve an acceptable safety and lifetime of the battery. However, also for other types of

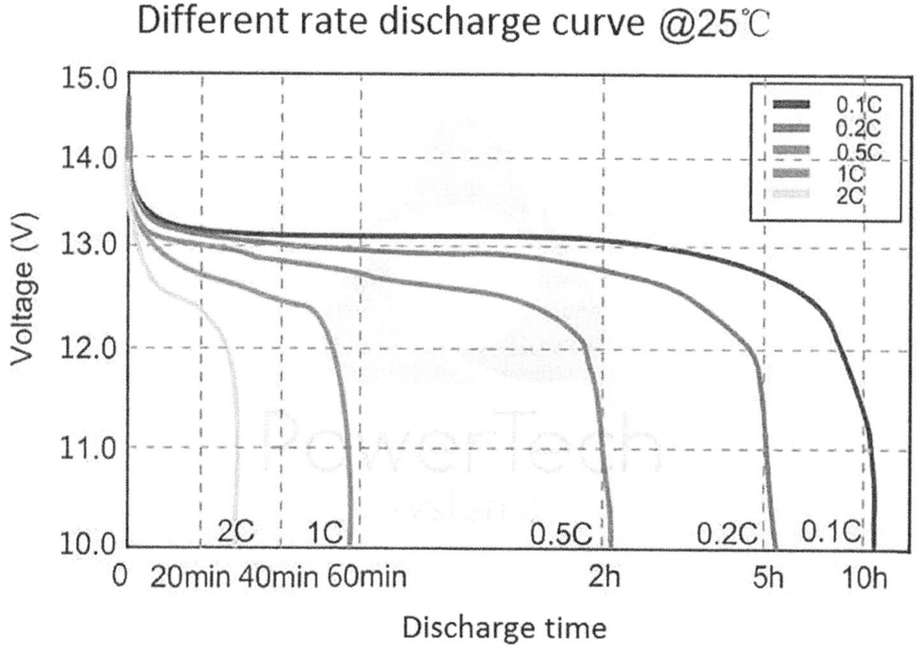

Figure 3.7 Discharge curve for LFP 12-volt battery PowerBrick+ (shown in Figure 3.6), according to PowerTech Systems. More data and charts are available for this type of battery [11]. Reprinted with permission.

batteries that don't require a BMS, such as lead-acid, a BMS can be a benefit, for example to prolong lifetime with passive cell balancing.

3.6. Battery monitoring

When planning to measure on a battery, learn about the dangers beforehand so they can be avoided, and use instrumentation that is suitable for the purpose. An ordinary multimeter set to measure DC voltage will normally have high internal impedance and therefore will not short-circuit the battery. However, it can short the battery in other cases, for example if it is set to measure current.

Measuring the total voltage of a battery consisting of many cells can in a healthy battery sometimes give an indication of SOC, as shown in Figures 3.3 and 3.4. For example, in an off-grid system, it can be important to know how often the battery becomes fully charged, how deep it is discharged, etc. However, for a deeper understanding of battery health it is best if cell voltages can be measured. If the terminals of all cells (or at least all blocks) are available for manual measurement and the battery is in a "steady state" situation for some time (where the current does not change, e.g. at rest), a multimeter can be used to inspect the individual cell (block) voltages. In this case, the difference between the cell (block) voltages is of most importance, because it is desired to keep all cells of a battery as similar as possible (to keep them at the same SOC levels).

One method to evaluate this is to make a table, where at regular intervals the voltage of each cell (block) is recorded, the average cell (block) voltage at that time is calculated and then the deviation of each cell (block) voltage from that average. (An example spreadsheet for this is available – see the website links listed later.) If the deviations from average are small, no action needs to be taken. If a problem is found at an early stage, it may be possible to correct, perhaps by an equalization charge. Another way could be to charge each cell individually, if possible. For lead-acid batteries a deviation in cell voltage of more than 0.1 V from average cell voltage at that time shows that something is wrong.

Especially for critical applications, such as uninterruptible power supply (UPS) systems for very important loads, continuous monitoring is better. In many cases, a BMS can provide monitoring, but remember that a BMS is involved in the control of the battery – thus, if there is, for example, a measurement fault in the BMS that may cause the BMS to control the battery in the wrong way while there is no reliable monitoring. So for some applications, an independent monitoring system may be required, where the monitoring does not interfere with the battery.

However, connecting a measurement system to a battery can be dangerous and increase the risk for faults, if done the wrong way. Measuring leads need to be well protected. In some cases protection is done by fuses on the measuring leads, but while a fuse may protect the wire, it might not be enough to protect a human. Another risk with fuses is that with batteries that can emit hydrogen, there may be an explosion risk if the fuse is place too close to the battery (but on the other hand you want the protection as close to the battery as possible for other safety reasons).

TEROC has developed a free, open-source software for battery monitoring, called Thor Battery Monitor (TBM). (See website links listed later to download.) It is, like established commercial products such as Batscan, instead based on placing protective resistors on the measuring leads, next to the battery terminals. These resistors are designed to limit the short-circuit current through these wires to a safe level below 1 mA even during fault when the full battery bank or charging voltage is applied, and prevent sparks, etc. As shown in Figure 3.8, these resistors can also be part of a voltage divider.

Voltage divider 1:95, including protective resistors, intended for charging voltage<360 V

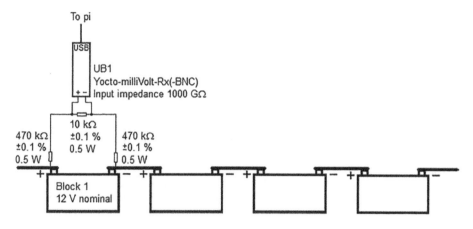

Figure 3.8 Example of measurement circuit for a 12-volt battery block. In a battery-monitoring system there would ideally be one such measurement circuit for each cell. Second best would be one for each block. A third and least expensive alternative is to monitor the midpoint voltage of the battery bank compared with the total (a fault indication is if the midpoint drifts too much from half the total voltage, but that can miss some faults).

For this to work well, an analog-to-digital converter with extremely high input impedance has been selected (which enables even higher resistance values than in Figure 3.8). The TBM software, which collects and processes the data, runs, for example, on a Raspberry pi. TBM can send the data to another software, Emoncms, for logging and web visualization. The compressed file that can be downloaded for TBM contains an add-on to Emoncms to enable real-time comparison among blocks or cells with a bar chart. Logged data can be used to see how voltages of cells or blocks change over time, for example during discharge, which is another way to identify abnormal cells or blocks. TBM also counts amp-hours and, based on that, estimates SOC. A plan is to add Modbus support in the future to communicate measurements to a higher-level system, described later in the book. TBM could also be extended to measure, for example, renewable energy input current and load current, something already done in other products such as PentaMetric.

3.7. Internet websites

Independent testing of batteries of interest for renewable energy applications under typical Australian conditions, mainly lithium-ion batteries, batterytestcentre.com.au

Example spreadsheet for battery bank measurements, by TEROC, www.teroc.se/web/page.aspx?refid=70

TBM – Thor Battery Monitor, free open-source software for monitoring of battery bank, by TEROC, www.teroc.se/web/page.aspx?refid=77

3.8. References

[1] J. K. Kaldellis (ed), *Stand-Alone and Hybrid Wind Energy Systems*, Woodhead Publishing/CRC Press, 2010.

[2] Sunfire, "Sunfire produces sustainable crude oil alternative". [Online]. Available from: https://www.sunfire.de/en/company/news/detail/sunfire-produces-sustainable-crude-oil-alternative [Accessed 2019-02-19].

[3] Chalmers, "Syntetisk bensin viktig pusselbit i fossilfri fordonsflotta". [Online]. Available from: https://www.chalmers.se/sv/institutioner/see/nyheter/Sidor/Syntetisk-bensin-viktig-pusselbit-i-fossilfri-fordonsflotta.aspx [Accessed 2019-02-19].

[4] lygte-info.dk, "Batteries with ridiculous ratings". [Online]. Available from: https://lygte-info.dk/info/BatteriesRidiculousRatings%20UK.html [Accessed 2019-03-11].

[5] P. Gipe, *Wind Energy Basics: A Guide to Small and Micro Wind Systems*, White River Junction, VT: Chelsea Green Publishing, 1999.

[6] batteryuniversity.com, "BU-107: Comparison table of secondary batteries". [Online]. Available from: https://batteryuniversity.com/learn/article/secondary_batteries [Accessed 2019-03-16].

[7] ITACA, "A guide to lead-acid battries". [Online]. Available from: https://www.itaca.org/a-guide-to-lead-acid-battries/part-1-how-lead-acid-batteries-work/ [Accessed 2019-03-16].

[8] R. Perez, "Lead-acid battery state of charge vs. voltage", *Home Power Magazine*, no. 36, August/September, 1993.

[9] Stockholm University, "Swedish research multiplies the life of rechargeable NiMH batteries". [Online]. Available from: https://www.su.se/english/research/research-news/swedish-research-multiplies-the-life-of-rechargeable-nimh-batteries-1.418740 [Accessed 2019-03-09].

[10] F. Larsson, P. Andersson, P. Blomqvist, B-E. Mellander, "Toxic fluoride gas emissions from lithium-ion battery fires", *Scientific Reports*, 2017-08-30. [Online]. Available from: https://www.researchgate.net/publication/319368068_Toxic_fluoride_gas_emissions_from_lithium-ion_battery_fires [Accessed 2019-02-24].

[11] PowerTech Systems, "Lithium-ion battery 12V – 100Ah – 1.28kWh – PowerBrick+". [Online]. Available from: https://www.powertechsystems.eu/home/products/12v-lithium-battery-pack-powerbrick/100ah-12v-lithium-ion-battery-pack-powerbrick/ [Accessed 2019-03-17].

[12] M. Murnane, A. Ghazel, "A closer look at state of charge (SOC) and state of health (SOH) estimation techniques for batteries". [Online]. Available from: https://www.analog.com/media/en/technical-documentation/technical-articles/A-Closer-Look-at-State-Of-Charge-and-State-Health-Estimation-Techniques-. . . .pdf [Accessed 2019-03-17].

Chapter 4

Use of energy and electricity

Energy consumption in the world in 2017 amounted to a total of 157,135 TWh. Consumption grew 2.2% during that year. That was an increase compared with the 10-year average increase of 1.7% per year. Regarding the sources to cover this use of energy, natural gas accounted for the greatest increase, followed by renewables and oil. In China the energy consumption rose by 3.1%. China had the largest growth for energy use for the seventeenth consecutive year.

The largest energy source in 2016 was oil, followed by coal and natural gas (Fig. 4.1). Still, fossil energy covered almost 80% of energy consumption, and the increase in energy needs was not covered by renewable energy. But we are starting to approach that situation. During the past decade the growth of modern renewables has increased, with 5.4% a year in average.

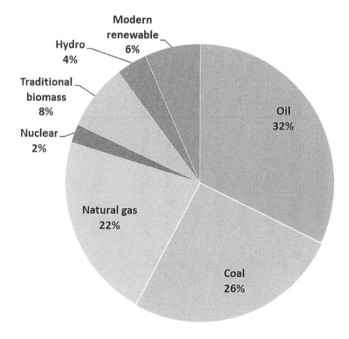

Figure 4.1 Final energy consumption in the world in 2016.

Source: REN21 [1]

Renewables contributed with 18.2% of energy consumption. For heating, traditional biomass contributed with 7.2%, and modern renewables (biomass/solar/geothermal heat) 4.1%. For electricity, hydropower generated 3.7%, and wind/solar/biomass/geothermal/ocean power 1.7%. In addition, 0.9% biofuels were used for transports.

The consumption of energy varies significantly among countries. Figure 4.2 gives some examples. Three Nordic countries – Denmark, Norway and Finland – have about the same population, but energy use varies much. Norway and Finland use much more energy than Denmark.

The significant difference can have several reasons. Norway and Finland have historically had great potential to develop electricity production early with hydropower, resulting in low electricity prices. Denmark has lacked hydropower but, since the 1980s, wind power has been built that today meets more than 40% of the electricity demand. But now electricity prices without taxes are determined on the common electricity market for the Nordic countries, and the price differences are not as big for the trade price.

However, Denmark has, through general taxes on electricity and a special carbon dioxide tax on fossil power generation, significantly higher electricity prices for consumers than Finland and Norway. The price is about double.

Denmark has also decided to run an efficiency policy package. The most important element is "The Danish Energy Efficiency Obligation" (EEO), or "White Certificates," which stimulate energy efficiency. Denmark is an example of how governmental decisions can significantly affect the use of energy. But even in Denmark, consumption exceeds the world average by more than 70%.

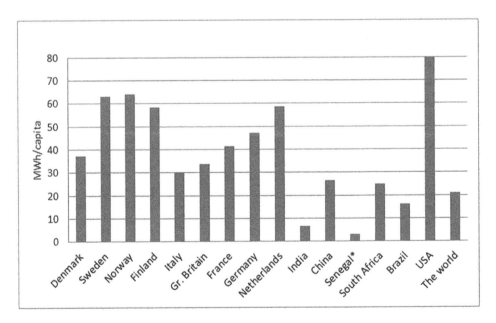

Figure 4.2 Energy consumption per capita in some countries in 2017.

Note: *Data for Senegal is from 2014.

Source: BP [2] and the World Bank [3].

Energy consumption per capita of Norway, Sweden and Finland is also considerably higher than in other European countries such as Italy, France, Germany and the UK. Those countries also vary significantly in their consumption.

African countries use less energy than in Europe, but even there, the differences are high. In South Africa, the use of energy is eight times higher than in Senegal per capita.

The difference is also significant between the two largest countries in Asia: China and India. China's energy use is more than four times as high as India's per capita. However, the situation is not surprising given China's prolonged major industrial growth.

4.1. Energy use in different sectors

A breakdown of energy use per sectors is shown in Figure 4.3. It illustrates what energy is used for in 19 IEA (International Energy Agency) countries. These are highly developed countries.

It is interesting that in the transport sectors, the most energy is used for human transport, 21%. That is significantly more than for freight transport, 14%. If transports are made more energy efficient, it is obviously most important to find efficient transport alternatives for people. Transportation systems driven by electricity can contribute to this. For example, electric cars use about one fifth of the energy that a petrol car uses.

In the residential sector, representing 20% of energy demand, half of the use is for space heating (that varies significantly among the countries).

The services sector contains shops and department stores, schools, health care and other social activities.

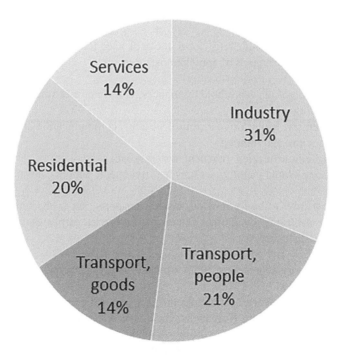

Figure 4.3 Percentage of energy use in different sectors in 19 IEA countries [4].

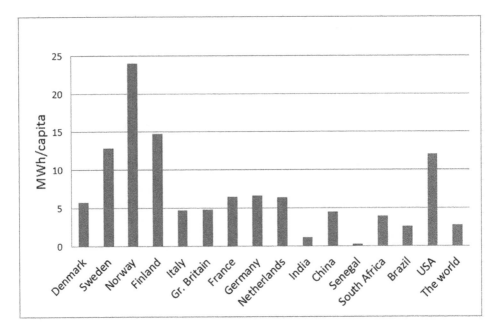

Figure 4.4 Average electricity consumption per capita in different countries in 2016 [5].

4.2. Electrical power

Electrical power is a very useful energy form. It can be used for almost all our energy needs. The following are some examples of applications:

- In households: Lighting, household appliances, consumer electronics and, sometimes, heating or cooling
- Within the industry: Motor drives, control systems, communication and other electronics, heating/cooling and lighting
- Trade: Refrigeration/freezing, payment systems, banks
- Public services: Water supply, wastewater treatment, schools, hospitals and street lighting
- Transports: Mainly in the form of rail-based transports, but in the future increasingly important for electric cars and other vehicles; even electric airplanes can be used, especially for short distances

In 1973, electric energy represented 8.6% of total energy consumption. This figure had increased to 16.3% in 2017.

Electricity consumption varies even more than for the total energy among countries. Figure 4.4 provides some examples. Norway has the highest electricity consumption per capita. Thanks to its potential to build many hydropower plants, electricity prices are very low there. In Norway, much electricity is used to heat buildings.

In developing countries electricity use is low. The expansion of electricity networks is also far from comprehensive.

Electric energy gets an increasing and important role and the share of electricity compared with total energy use can be expected to continue to increase. Often, electrification can contribute to a decrease in the society's emissions of carbon dioxide. Nowadays electricity use is influenced, for example, by charging of more electric vehicles, new industrial processes that replace the use of fossil fuel with electricity and establishment of data centers for internet services.

An example of new industrial processes is that the steel manufacturer SSAB in Sweden has begun to build a pilot plant for fossil-free steel production. The goal is to develop a fossil-free steel production using hydrogen instead of coke and coal. The new goal is to gradually move towards fossil-free steel production and eliminate carbon dioxide emissions from fossil fuels. By 2025 the goal is to achieve 10 million tons of annual carbon dioxide savings, and by 2045 the company expects to be completely fossil free. The investments are said to demand that new fossil-free power generation and power lines be built.

Figure 4.5 Compact fluorescent lamps (CFLs) are curly versions of the long-tube fluorescent lights. They are more energy effective than incandescent lamps, but only some are dimmable. They contain a small amount of mercury, so they are especially important to dispose of properly for recycling at the end of their lifespan.

Source: US Department of Energy

Figure 4.6 Light emitting diodes (LEDs) are the most energy-efficient lamps today and normally last 15–25 times longer than traditional incandescent lamps. LED lamps come in a wide variety, not just the bulb shapes shown here. Some are dimmable or offer features such as daylight and motion sensors. There are LEDs that work well outdoors, also in cold environments.

Source: US Department of Energy

However, the potential for more efficient use of energy should not be underestimated. Measures such as effective solutions based on electricity result in most cases in a significant energy efficiency improvement (Figs. 4.5 and 4.6). It is interesting to see that a country like Sweden has not increased the total use of electricity during the past years, despite a growing economy, and that electricity is used for more applications than before.

To overcome bottlenecks in the electricity system, there is also a largely untapped potential in energy demand management, also known as demand-side management (DSM) or demand-side response (DSR). That means shifting demand, by various methods such as automatic load shedding, financial incentives or behavioral change through education, from periods of peak demand to other times.

Using local energy storage, for example in grid-connected battery systems in combination with on-site generation of renewable energy, can also be a way to overcome bottlenecks in the electricity system. Small-scale solutions are typically faster to deploy than the traditional way of building new large, central power plants and power lines with ever higher capacities.

For small, independent electricity systems, where the cost per kW and kWh can be high, smart use of energy (including high energy efficiency) is especially important. It is usually much less expensive than increasing generation or storage capacity and can immediately reduce the risk of power and energy shortage.

Hydrogen production is also an option for storing electricity. When the proportion of weather-dependent power generation increases, hydrogen production can be the solution for handling overproduction and leveling in the systems.

4.3.   How we use domestic electricity

One important area for electricity use is the residential sector – how we use electricity in our homes. Table 4.1 shows some estimated values for the sector in Sweden [6] and some statistical values for the United States [7].

With more computers, battery chargers for smart phones, lighting and household electrical appliances, the number of electricity-consuming devices has increased sharply in recent

Table 4.1 Domestic energy use per year in Sweden and USA

Domestic electricity use in Sweden [6]		Domestic electricity use in USA [7]	
Average house/apartment	kWh	US residential utility customer	kWh
Clothes washers/dryers	750	Cooling/air conditioning	1600
Dishwashers	275	Water/space heating	1870
Electric equipment	625	Clothes washers, dryers (excl heating)	480
Refrigerator/freezer	1000	Dishwashers	210
Cooking	600	TV, computers, related equipment	840
Lighting	750	Refrigerator/freezer	920
Total	**4000**	Cooking	240
		Lighting	980
		Other miscellaneous use	3260
		Total	**10,400**

decades. But the increase in electricity consumption has not been so great. This is because many appliances are used for very short periods and, especially for home appliances and lighting, energy-efficient alternatives have been developed.

The figures for the USA include some heating that is missing in the Swedish figures. If this is eliminated and also the US-wide use of cooling/air conditioning that is insignificant in Sweden, the use is around 70% higher in the United States.

4.4. Energy labeling and ecodesign

The energy labeling system in the EU has been a great help for consumers to choose energy-efficient products. According to the EU Energy Framework Directive (2017/1369), the requirements for energy labeling are laid down in a process coordinated by the European Commission.

Products are currently (2018) marked with a scale from A+++ (most effective) to G (least effective). However, more energy-efficient products are coming on the market, and the scale will gradually be replaced with a new simpler A-G scale.

From 2021 onward, five product groups (refrigerators, dishwashers, washing machines, TVs and lamps) will be labeled with the new scale. Thus, a product labeled A+++ energy efficiency class today can become a B-class without changing energy consumption. The new A-class will initially be empty to leave room for future development. The changed scale will allow consumers to better distinguish between the most energy-efficient products.

Another important measure for more efficient energy use is eco-design requirements. EU legislation on this has been an effective tool for improving the energy efficiency of products. It eliminates the products with high energy consumption from the market, which helps to achieve the EU's energy-efficiency goals.

The Eco-design Directive contains two types of requirements:

The *specific requirements* specify exact limits for, for example, maximum power consumption or minimum amount of recycled material to be used in production.

Table 4.2 Residential energy use per year in an energy-efficient household

Energy efficient household		
	kWh	Energy label
Clothes washer	230	A+++
Clothes dryer	160	A+++
Dishwasher	240	A+++
TV, 50 inches	80	A+
Electrical appliances	200	
Fridge	100	A++
Freezer	200	A+++
Cooking	400	
Lighting	130	8×9W LED
Standby	40	10×0.5W
Total	**1780**	

The *general requirements* mean that:

- The product must be energy efficient and recyclable.
- Information must be given on how the product should be used and maintained. This will minimize environmental impact.
- A life cycle analysis for the product must be done to investigate alternative design solutions and improvement opportunities.

Products that do not meet the requirements cannot be sold in EU countries. An example is incandescent lamps that have been phased out since 2009 and now also halogen lamps. Another important rule is that standby consumption (consumption when a product is not used) for many products may not exceed 0.5 watts. Many products must also switch to a power save mode after the shortest possible time when not in use. Previously, electricity consumption could be significantly higher annually when, for example, a TV was in standby compared with when it was in operation.

The energy requirements of a household that uses products with the best energy data is shown in Table 4.2. This consumption of electricity compared with the Swedish data given in Table 4.1 is 55% lower. For the specific area "Lighting," use consists of eight 9-watt LED lamps. A 9-watt LED lamp gives about the same amount of light as an old 60-watt incandescent lamp, so the lighting ought to be just as good.

The energy savings between the Swedish domestic consumption shown in Table 4.1 and the "energy-efficient household" is 2220 kWh. That is what an ordinary electric car needs to be driven a yearly distance of more than 10,000 kilometers. Through these savings a family can avoid a lot of fossil fuel.

4.5. Examples of efficient use of energy in the commercial sector

Variable-speed motor drives (frequency converters) can in many applications reduce energy consumption of electric motors significantly. This also provides a soft start for motors (reducing in-rush current), which can be of special importance in small power systems, for example supplied by an inverter.

In many shops, energy can be used more efficiently while displaying products that must be kept cold, by using transparent doors/lids on refrigerators and freezers. It is important to avoid the common mistake of heating and cooling at the same time.

For data centers that require an uninterruptible power supply (UPS), there is potential to avoid unnecessary power conversion steps by using 380-volt DC to power servers and other loads designed for this voltage (Fig. 4.7). That can be of special importance for countries where waste heat is not useful. DC-UPS systems are normally built with the UPS battery directly on the DC-bus, which should lower the risk of failure, compared with an AC-UPS. It is also relatively simple to create a redundant DC-UPS, and solar PV can be integrated on the DC-link too.

Figure 4.7 John Åkerlund from NetPower Labs shows a DC-UPS for a data center.
Source: S. Ruin.

4.6. Internet websites

North America, Japan, Taiwan, Europe, www.energystar.gov
Australia and New Zealand, www.energyrating.gov.au

4.7. References

[1] REN21, "Renewables 2018 global status report". [Online]. Available from: http://www.ren21.net/ gsr-2018/ [Accessed 2019-05-30].
[2] *BP Statistical Review of World Energy*, 2018. [Online]. Available from: https://www.bp.com/en/ global/corporate/news-and-insights/reports-and-publications.html [Accessed 2019-05-30].
[3] The World Bank, "Indicators" (Climate Change, Energy Use). [Online]. Available from: https:// data.worldbank.org/indicator
[4] IEA, "Energy efficiency statistics", 2017. [Online]. Available from: https://www.iea.org/statistics/ efficiency/
[5] The World Bank, "Access to electricity (% of population)", 2016. [Online]. Available from: https:// data.worldbank.org/indicator/EG.ELC.ACCS.ZS
[6] E.ON, 2018. [Online]. Available from: https://www.eon.se/privat/for-hemmet/energiradgivning/ normalfoerbrukning.html
[7] EIA – US Energy Information Administration, "Frequently asked questions: how is electricity used in U.S. homes?", 2018. [Online]. Available from: https://www.eia.gov/tools/faqs/faq.php?id=96&t=3

Chapter 5

System design

5.1. Basic considerations

5.1.1. Grid connection vs. stand-alone power systems

As a traditional rule, the smaller a power system is, the larger the costs of electricity generation per kWh. Therefore, it has been a natural development in many countries to connect almost all producers and consumers by power lines, forming a national power grid. The national power grid is usually connected to the power grid of neighboring countries, thus forming an even larger interconnected electricity system. Grid connection of a power plant is sometimes also called *interconnecting with the utility* or *utility inter-tie*.

The historical trend towards grid connection has also been driven by subsidies and the fact that, in the past, technical solutions for stand-alone electricity production were not so well developed. Today this situation has changed, and much better technology is available for those without grid connection, but knowledge about this is not yet as widespread as it ought to be. Stand-alone power systems are still often overlooked as a possibility or poorly designed. Knowledge regarding the maintenance of such systems may also be limited.

Stand-alone power systems (Fig. 5.1ab) are also called *autonomous, off-grid* or *remote area power systems (RAPS)*. Sometimes the term *hybrid system* is also used, for power systems built as a combination of two or more generating technologies, which is often the case for off-grid or grid backup systems (a combination of grid-connected and off-grid). Hybrid systems are described later in the book. Other related terms are *micro-grid, mini-grid* and *island grid*, which refer to the isolated local power grid, which can function without connection to the national power grid. Such *isolated grids* can in some cases be quite large, and can be seen as an intermediate case between small off-grid systems and large central-grid systems.

Where the grid is available at a not too high cost, it usually makes sense to use it, even if stand-alone systems are getting better. Stand-alone systems based on sources like solar and wind typically have a large surplus generation during part of the year and sometimes much energy is wasted this way. The grid normally enables surplus generation to be sold and used by someone else, but sometimes poorly designed rules makes this difficult. However, it is a danger that our society is so dependent on the grid for almost everything to function.

Figure 5.1ab Example of the installation of a packaged power solution, in the form of a cabinet module, which is the indoor part of a stand-alone solar system (prepared to be expanded to a solar-wind hybrid system with grid backup capability). Western MPPT solar charge regulator in upper left-hand corner. Studer Innotec inverter (with grid-forming capability, i.e. to control grid voltage and frequency on the 230 V AC output below) on the right-hand side. Both are mounted on fireproof material. DC fuses in the middle (with fuses for both plus and minus, because the DC system is floating). In the battery compartment (with non-conductive surfaces) are four Sonnenschein SB6/200 batteries (lead-acid gel type, non-spillable) connected in series for 24-V DC system voltage. There are ventilation openings in the battery compartment and ventilation in the room. Transparent cabinet covers prevent unauthorized access. In this traditional system architecture, the batteries are directly on the DC-bus.

Small-scale power plants can be used in many circumstances, from grid connection to stand-alone systems. It is important to understand that there are some fundamental differences between the applications, such as:

- Control of voltage and frequency. When a small generator is grid connected, there are normally other power plants on-line that will control the voltage and frequency (power balance) of the grid. For example, electrical power to the grid must equal consumption at all times, which is in an AC grid done by regulating the frequency, and on the grid that is usually the responsibility of someone else. A stand-alone power system must, on the other hand, always keep the power balance. So for stand-alone applications you must see the local power *system* as a whole, not just look at individual power plants, and take responsibility, for example for power balance.
- Cost. Stand-alone power systems are generally more expensive to buy and operate than comparable grid-connected power plants. Therefore, stand-alone solutions are traditionally only used far from the central power grid, but this may be changing.
- Project size. A stand-alone project is often much smaller than a grid-connected project, but both could be complex. If the project is complex, it can be difficult to handle in a small project, where the budget is normally also small. Complexity can, for example, be technical (especially in the past, some hybrid systems were more like research and development projects). Packaged power solutions can reduce complexity for users.
- Parties involved. Grid-connected projects normally involve large companies, which, for example, often charge a standard fee for all their services, including repairs. Stand-alone systems are sometimes so small that they are made and maintained by private individuals, even with home-built equipment. What if the stand-alone system needs unplanned maintenance at an inconvenient time or a costly repair?

An example of a tool that was made for grid-extension planning is ViPOR (village power optimization model for renewables). ViPOR calculates which houses should be powered by isolated power systems and which should be included in a centralized distribution grid.

However, to determine the suitable system type is not only a matter of technical optimization. For example, to share a common system, a village can introduce many non-technical challenges regarding cooperation among the inhabitants, billing, preventing abuse of the system, etc.

5.1.2. Economy

Some common characteristics of many renewable energy technologies, compared with conventional technologies, are higher initial costs and lower operating costs, and that they are often cost effective on a life cycle basis.

Comparing the cost per installed capacity (e.g. cost/kW) is therefore often of very limited use, except when comparing generation technology with standardized ratings, such as one solar PV system to another. Usually the cost of energy (e.g. cost/kWh) is a much better measure. For renewable sources such as wind and solar, where the operation and maintenance costs are normally low compared with the initial investment, a practical figure for rough comparison of investments is the cost per yearly kWh. For example, if a particular 800-kW wind turbine would require a total investment of 12 million SEK and yearly generate 2 million kWh, the relative or specific investment would be 6 SEK per yearly kWh. A smaller wind turbine can cost, for example, 12 SEK per yearly kWh and a larger 3 SEK per yearly kWh.

It is important to understand that the size of the power plant in question can have considerable influence on the cost of energy. Some generation technologies like wind have an economy of scale, where the cost of energy drops remarkably when moving up in size from a small to a large wind turbine. Other technologies have a more linear relationship between size and cost – two PV modules double the power compared with one and cost about twice as much as one.

Thus, generating technologies with an economy of scale typically have an economical advantage for grid connection and for other large systems. However, in very small stand-alone systems the situation is the other way around.

To properly assess the economy of a project, the ideal is if the entire life cycle can be calculated, including project development, purchase cost of equipment, fuel costs, transportation, operation, maintenance, scrap value/restoration/recycling costs, etc. An example of a free tool available is RetScreen.

Especially for stand-alone systems, it is worth keeping in mind that the renewable sources are sometimes sized to cover a worst-case scenario, which means that they will be oversized for a normal situation and some power often has to be curtailed. Thus, some of the kilowatt hours that can be produced might not be used. This is taken into account by simulation software, described in the hybrid systems section.

Grid-connected small-scale renewable energy systems of today can in some cases do so much more than produce energy, and traditional calculation models don't include such other services. For example, the power electronics in Ferroamp's EnergyHub system (see Figure 3.5) provide a measurement and analysis system for the electricity of the facility and a function to automatically equalize currents among the phases in a three-phase system. Together with a battery, power peaks can also be reduced ("peak-shaving"). Such functions can help reduce the grid connection and subscription costs. In some cases it can even enable business, such as car charging, that would otherwise not be possible to connect to a weak grid. Add-on services for grid frequency regulation are coming, which can create an income (see "Municipal utility company using V2G and solar PV" in Chapter 6). And how do you value a grid backup functionality?

5.1.3. Security of supply

Security of supply is affected by a number of factors. One thing that sets stand-alone renewable energy systems apart from conventional grid connection and engine-driven generators operating on fossil fuel is that they don't rely in the same way on the society's infrastructure. This can be of vital importance in the case of events such as a major disaster or fuel embargo.

Other important factors related to security are discussed later in this chapter, including operational and maintenance issues. Despite all the opportunities brought by small-scale renewable energy systems, there are many things to consider, especially where a high security of supply is desired. If, for example, a stand-alone system is poorly designed/operated/maintained, it can provide an unacceptably low security of supply. Such issues have in some cases caused islands to go back to 100% fossil diesel use, after trying renewable energy for a while.

5.1.4. Selection of operating voltage

For traditional grid-connected systems, the selection of operating voltage for the most small-scale renewable energy systems is simple – just use the voltage of the grid (perhaps you have a choice between single-phase and three-phase). There are actually other choices available, based on DC, not least for off-grid. To use a common DC bus is especially interesting for combining solar PV, batteries, fuel cells and variable-speed generators or motors, because they are all based on DC.

As explained in the case study "House Unplugged" in Chapter 6, 760 V DC (±380 V DC with a grounded DC neutral) is a voltage used in some cases to build a DC nanogrid, where, for example, charging of electric cars can be integrated. A voltage level of several hundred volts is practical when there is a need to transfer much power on the DC-bus and in certain cases also because some three-phase motor drives can be connected to the bus (e.g. for uninterruptible motor drive applications).

Emerge Alliance has defined 380 V DC as a standard for data/telecom centers. As mentioned in the section "Examples of efficient use of energy in the commercial sector" in Chapter 4, two of the reasons for using DC in such applications are efficiency and reliability. Special plugs and receptacles are available for these applications (Fig. 5.2). This voltage is also used in some DC nanogrids for other uses. Renewable can be integrated on a 380-volt DC-bus as well, for example the wind turbine T701 from Pika Energy (see Table 2.2). It should not be too difficult to approve more equipment for such applications, because some loads originally designed for 230 V AC can actually work also with around 350 V DC.

Figure 5.2 Example of special plug and receptacle for 380-volt DC.

Source: S. Ruin.

In other cases, safety is a driver for lower voltages. An established voltage in telecom is −48 V DC, where the positive side is grounded (0 V). India has a 48-volt DC standard, for which, for example the company Cygni sells products. Emerge Alliance has defined 24 V DC as a standard for occupied space. 12 V DC is common in vehicles. Based on such standards, small off-grid and in some cases grid backup systems can be built safer and often cheaper and more efficiently without inverters, when all equipment (loads, solar PV, etc) is connected to the extra-low voltage DC. This could save many lives, especially where people are unaware of the dangers of electricity, where there are few qualified electricians or where there is no culture of making safe electrical installations.

Another standard worth mentioning here is USB, which normally works with 5 V DC. There are small solar home systems today that use the USB standard, for example to charge mobile phones. With the new USB-C, higher power can be transferred. Up to 100 W at 20 V DC is possible with the USB power delivery specification.

The downside of using extra-low voltage is, of course, the power limitations, because for a given power the current will be higher for a low voltage. That can be overcome by using a higher AC or DC voltage where needed.

5.1.5. Logistics

One of the uses of small renewable systems is powering remote locations, where transporting large or heavy equipment to site can be difficult (Fig. 5.3). Getting spare parts there can be another issue. In such cases, planning the logistics is especially important. This can also be a driver for using smaller generators than would otherwise be optimal. For example, standardizing on a few types of components and/or considering the use of locally made equipment can have logistical and other advantages for a rural electrification program.

Figure 5.3 PowerSpout Pelton turbines carried in Nepal.

Source: Rural Integrated Development Services (RIDS).

5.1.6. Mobility

In the context of small-scale renewable energy, mobility can have different meanings:

- The smallest systems can sometimes be carried. For example, Sundaya systems have a lightweight lithium battery integrated in an energy-efficient LED lamp, which can be used as a flashlight.
- There are larger systems that can be transported when needed, such as in a container.
- For example, on a camper van, a renewable energy system may be needed to travel with the vehicle. In some cases, such as electric boats, the on-board renewable energy system could even drive the propulsion.
- Stationary renewable energy systems can in some cases charge electric vehicles.
- Some electric vehicles can be used, for example, for vehicle-to-home (V2H) or vehicle-to-grid (V2G).
- There are electric vehicles, such as Zbee, where batteries can be swapped and also used for different purposes.

This opens up a world of new opportunities – and challenges – for small-scale renewable energy systems.

Be aware that on boats and similar applications, issues such as galvanic corrosion are of special importance. There are special books on electrical systems for boats.

5.1.7. Some electrical considerations

5.1.7.1. Fuse selectivity

A main fuse and a branch fuse are said to be selective if the branch fuse will trip, in case of a fault or overload in that part of the circuit, before the main fuse opens. The same goes for MCBs, which are often used instead of traditional fuses. Selectivity is desirable, because otherwise, for example, an entire mini-grid could have a blackout whenever a fault/overload occurs.

However, the inverters used in many off-grid systems can normally only supply a very limited short-circuit current and have a very quick over-current protection built in, which means that the inverter will often trip before any fuses/MCBs in the AC distribution will trip.

Some DC/DC converters can also have problems delivering short-circuit current, but they might during overcurrent reduce voltage rather than trip. Electronic fuses (E-fuses) are integrated circuits that might help to achieve selectivity.

5.1.7.2. Grounding, ground-faults and arcing

Proper grounding, for example of metallic enclosures, is sometimes an overlooked issue. The *Mini-grid Design Manual* provides some guidance on, for example, how to install a grounding rod, but it doesn't cover much regarding small-scale renewable energy systems, because it was written when such systems were at an early stage [1].

Note that there can be different requirements on grounding, ground-fault protection and arc protection in different countries. In any case, it is recommended to pay close attention to detail when wiring, for example, at a PV array where wires are installed in free air and are more easily damaged. The higher the voltage, the higher is the risk of, for example arcing, especially in a DC PV array which can cause an arc that keeps burning for a long time.

5.1.7.3. Electromagnetic pulse

An electromagnetic pulse (EMP) could be caused, for example, by a solar storm or nuclear explosion. It could damage, for example, electronic equipment, including power electronics and control systems in small-scale renewable energy systems. EMP testing or protection is probably an area that needs more attention, especially for systems that are intended to function during an emergency.

5.2. Hybrid systems

Especially when designing the energy supply for an off-grid or isolated grid application, it is important to consider using a hybrid system, which is a combination of two or more generating technologies (of which at least one is usually renewable). When made properly, this combination can often overcome the limitations associated with systems using only one generating technology. Under many conditions, in particular on remote locations far from the main power grid where a diesel power system is the conventional generating technology, hybrids can provide the lowest cost of energy. An example of a hybrid system is to combine a wind turbine and a diesel genset to form a wind-diesel system.

Places that have an unreliable grid connection and need a grid backup can also benefit from this technology. In such cases, hybrid power systems can be built to function like a UPS combined with on-site renewable energy generation.

So instead of first selecting a particular generating technology for the site in question, it is often a better idea to consider hybrid systems from the beginning in the design process.

In a way, all the power plants connected to a national power grid function together like a giant hybrid system. However, the interactions between the power plants do not have to be considered much when connecting a relatively small power plant, for example a wind turbine, to such a grid. Therefore, the hybrid system insight is not needed for grid connection of relatively small power plants, unless there are so many of them that they together play a significant role in the power grid.

5.2.1. Hybrid systems for electricity supply

A hybrid power system can be built to supply a few light bulbs in a house or to supply the electricity of an entire town. Naturally, different sizes of power systems will have different characteristics, as shown in Table 5.1. Aggregation of many households and other types of electricity consumers in, for example, a village or a town provides a positive impact on power system operation. A town typically also has a more efficient diesel power system.

Table 5.1 Generalized characteristics of different sizes of off-grid or isolated grid power systems (approximate numbers)

Power system size	Typical characteristics of conventional power system				Typical amount of renewable energy in hybrid systems	Typical minimum electricity storage capacity in hybrid systems
	Peak/ average load ratio	Diesel power system	Fuel consumption at no load (compared with full-load consumption)	Relative cost of generation		
Household	Up to 100	Single genset	25%	Extremely high	Very high	Days
Village	4	Single genset	10%	High	High	Minutes
Town	2.5	Multiple gensets	5%	Medium	Low-medium	None

5.2.1.1. Household-size and below

Let's start by looking at the typical loads for a single household in Sweden, which does not use electric heating or an electric vehicle. The average load is typically 100 to 500 W, while the peak demand is approximately 10 kW if an electric stove is used. Thus, the ratio between peak and average load is 20 to 100 for this example. If the household were supplied only by a diesel genset (or any type of genset), sized to cover the peak demand of 10 kW, it would normally run very inefficiently because of the relatively high fuel consumption at low load. The resulting costs of power generation are extremely high, often in the order of € 1/kWh. Households at remote locations, which use this kind of electricity supply, can normally only afford to operate it a few hours a day. The load situation would improve if another form of stove is used, so the peak demand could be lowered and a smaller genset could be used.

It is normally optimal for off-grid household-size systems to cover most of the electricity consumption by renewable energy and to have a battery bank for energy storage. This can significantly lower the cost of energy, and additionally, power can be provided all the time. Wind and solar photovoltaics (PV) are often combined because they complement each other on a daily and seasonal basis. A weak spot has often been the batteries but, as explained in Chapter 3, there are solutions to that.

Traditional small, battery-based hybrid systems are well explained in literature, for example in Paul Gipe's *Wind Power*. Often, the DC output of the wind turbine's charge controller can simply be connected in parallel with the charge controller of a PV array. Based on the point of load sharing between the different sources, such a system is usually called a DC-bus system, although the loads could use DC and/or AC. In such an off-grid scenario, AC loads would require a grid-forming (master) inverter, which converts DC to AC and controls AC voltage and frequency on the isolated AC line, which could be a micro- or mini-grid.

Some inverters for this kind of applications have many more functions, often including the ability to work bi-directionally, and can then better be designated DC/AC converters. This can allow for mixed architectures, where some generation components are connected to the DC side and some to the AC side. Regarding connection of AC-generating units, there are two main principles, which can in some cases be combined.

One principle is that the converter has an AC input, which the converter can switch on and off, and this normally is used to connect a genset (that controls AC voltage and frequency). When power is detected by the converter on such an AC input, the converter is usually designed to synchronize its AC output with the input, before the input is automatically connected to the output. Then the converter will typically change operating mode to charge the battery bank, thus acting as a charger (possibly in parallel to other generating units on the DC-bus), while the genset has taken over the control of AC voltage and frequency. In an IEA report [2], this architecture is called "single switched master mini-grid," because grid forming control is switched between the genset and the battery converter.

The other principle is that generating units operating as grid-following "slaves" are connected to the AC output side of the converter, where they act as negative loads and in some cases it is even possible for them to charge the battery bank on the DC-bus, through the converter while it is still controlling AC voltage and frequency. To avoid overcharging of the battery bank, one method is that the master converter can temporarily raise the AC frequency a little. It could be a fixed higher frequency, which should cause the "slave generators" to stop. A smoother control can be achieved by stepless droop control of the frequency ("frequency shift"), which is used in the following approach.

One approach for battery-based systems is that all units for generation and storage are connected via an AC-bus. It can be used both for household-size systems and larger systems, such as villages. AC-bus configuration has special advantages if the different components in the system must be spread out, for example if PV arrays are to be placed on different buildings, already connected on an AC mini-grid. With this type of equipment it is also convenient to convert, for example, existing grid-connected PV systems to grid backup systems with renewable on-site generation (e.g. to supply critical loads in a building in case of a central power outage). However, AC-bus systems can be more expensive and complex than DC-bus systems. In addition, AC-bus systems can have lower efficiency, for example in a situation where the generation mainly takes place during the day and consumption mainly at night, because the power has to go through more conversion stages when charging the battery bank. Fortunately, the battery banks of several hundred volts that are becoming more common today allow for transformer-less battery inverters to be used with higher efficiency and lower price, compared with heavier inverters for extra-low voltage.

Regardless of the configuration used (DC-bus, AC-bus or mixed AC/DC-bus), the system design in household-size systems is normally made for the genset(s) to operate very little, if any genset is used at all. The reason is cost, as illustrated in Table 5.1.

5.2.1.2. Above household-size

Power systems above household-size will often have a more beneficial peak/average load ratio and more efficient diesel power systems with lower relative fuel consumption at low load. Consequently, the cost of energy will be lower. In such systems, an energy storage for long autonomy might not be as economical as in a household system, and it is probably not optimal to oversize the installed renewable energy generation too much. The amount of "uncontrollable" generation, such as wind, is a decisive factor for the system design.

For example, in wind-diesel systems the amount of wind power is often called *wind penetration*, and can be defined based on power or energy. Low wind penetration does not require complex technology, and can often simply be accomplished by connecting all power plants to an isolated AC grid, if power quality and safety issues are observed. When the wind power production is always less than the load, and other power plants are constantly on line to control grid frequency and voltage, the wind power saves fuel by reducing the load on other power plants. This is similar to connecting a wind turbine to a large national grid. The disadvantage is that it does not save as much fuel, especially if an unsuitable type of genset is used.

In a similar way, the more general term *renewable penetration* can be used to describe the amount of renewable energy in a hybrid system, regardless of which renewable source it is (Table 5.2). Normally, higher renewable penetration means technical challenges, but up to a certain point they are worth doing.

Another decisive factor is that most gensets require a certain minimum load, as explained earlier in Chapter 2. The low-load characteristics are of importance especially for some systems, where operation at low load can often not be avoided.

For a town with an isolated power system, a hybrid system might consist of a multiple diesel power station complemented by a few wind turbines, where the wind covers, say, 30% of the annual electricity consumption. Such an example is shown schematically in Figure 5.4. This type of system could also be described as a "multi master rotating machine dominated mini-grid architecture," to use the words of the IEA report [2]. Here, the grid is formed by a diesel power plant consisting of two or more gensets, with at least one of them operating continuously, thus controlling grid voltage and frequency. Normally, each genset used in this architecture is equipped for parallel operation. This includes synchronization before connection and load sharing. A supervisory control unit is normally used for dispatch functions such as to control load transfer from one unit to another and keep a sufficient spinning reserve.

Note that gensets are almost always equipped with synchronous generators, as shown in Figure 5.4, which are able to control grid voltage. Synchronous condensers and special gensets where the generator can be temporarily disconnected from the engine by a clutch have the same ability, but do not require the engine to be running to perform voltage control. Grid frequency can be controlled by the engine's speed governor, which normally reacts fast enough to cope with fluctuations in wind and load.

In the example of Figure 5.4, only wind turbines are shown, but solar generation could be used in a similar fashion, connected to the town grid in much the same way as to a large grid, using grid-following "slave" inverters. There are systems, such as CloudCAM, for short-term prediction of solar generation in order to avoid instability from sudden drops in solar generation (e.g. by solar sites pre-emptively ramping down with a limited ramp-rate).

Normally it is a benefit in such systems if the wind turbines have a relatively smooth output, so the spinning reserve of the diesel power station can be reduced. Using wind turbines with variable speed and pitch control (as in Figure 5.4) can help smooth the power output. Using many smaller, geographically distributed wind turbines or other forms of renewable

Table 5.2 Classification of renewable penetration (contribution) levels [2]

Class	Penetration	
	Peak instantaneous (power)	Annual average (energy)
Low	<50%	<20%
Medium	50–100%	20–50%
High	100–400%	50–150%

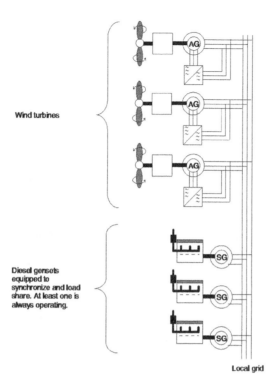

Wind turbines

Diesel gensets
equipped to
synchronize and load
share. At least one is
always operating.

Local grid

Figure 5.4 Schematic example of a wind-diesel system for a town.

electrical generation instead of a single large unit can also be of benefit for this purpose. Some wind turbines, such as XANT, are designed for microgrid operation.

Special care may have to be taken, to ensure, for example, that:

- Wind turbines can cope with frequency and voltage variations, which may be larger on isolated grids than on ordinary, national power grids.
- Diesel gensets are given adequate time to start, for example in case a wind turbine shuts down because of high wind speed (no cut-out wind speed or gradual cut-out can be a benefit).
- Wind power production is not too high (some wind turbines can temporarily limit the turbine power output, regardless of wind speed, to avoid this).
- Power quality and reactive power issues are considered.

With an increased renewable penetration level, the fluctuating and intermittent characteristics of wind and solar generation normally make grid stability issues a primary concern. A secondary concern is optimizing the contribution of all generation sources. Modern power electronics have enabled a high level of controllability, which is very useful for this purpose.

If more renewable generation, for example by solar PV, is added to the example shown in Figure 5.4, then a point may be reached when the renewables can, during short periods, cover more than 100% of the present power consumption. While this could be handled, for example, by an automatic control system to temporarily curtail the renewable output when genset

Figure 5.5 Grid-following Viessmann/SMA inverter for solar PV, with DC and AC discon-
nects (in a hybrid system further explained in the case study "Home with seasonal storage of hydrogen" in Chapter 6).

Source: G. Sidén.

load is low and/or to control load (e.g. to turn on heating loads or "dump loads" to use the
renewable energy), another way is to install one or more battery inverters with similar grid-
forming capability as a genset. The battery inverter can handle reverse power situations and
store excess generation for later use. This is actually the same as the AC-bus configuration
described earlier for household size, but larger and typically including several battery inverters. In the IEA report [2] that is called a "multi-master inverter dominated mini-grid," and
enables large systems to be built where there may not even be any genset operating. Electric
vehicles could also participate in frequency regulation of the AC-bus.

Another architecture, which also enables the genset to be shut down, is shown in Figure
5.6. Here, all the power normally flows through a single grid-forming inverter. It is a relatively simple integrated system, where special variable-speed wind turbines with suitable
characteristics are used, which feed the common DC link. Unlike smaller systems, the DC
link is here operating at several hundred volts DC and the battery bank can be optional.
The same DC link is fed by the rectified output of the genset(s). This can enable very high
renewable penetration levels and enables variable-speed genset(s) to be used for even larger
fuel savings. Such systems have been built with industrial motor drives, which are available
in a wide variety of sizes and normally cost less than other corresponding power converters
(but need special filter, etc., for this application). All this contributes to keeping costs down.
However, this architecture may not be the most suitable in case the generation sources are to
be spread out over a large area with an existing AC grid in good condition.

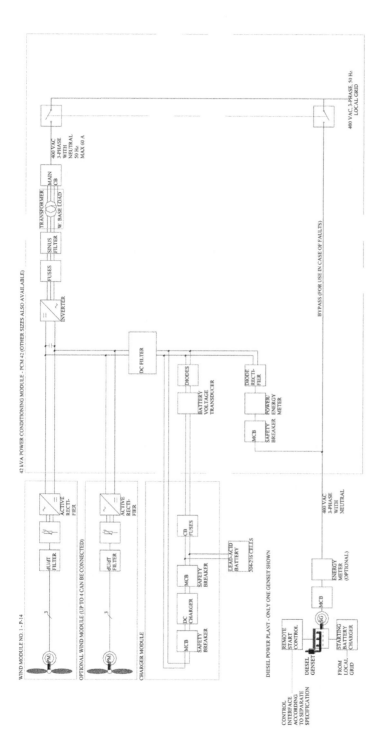

Figure 5.6 Example of a single master mini-grid, an architecture which has demonstrated that wind power can during windy periods be the only energy source of an island, controlling grid voltage and frequency, even with the battery bank disconnected and genset turned off. In this example, the voltage of the DC-bus can be considerably higher than the battery bank voltage, because they are not directly connected.

High renewable penetration is usually most economical, especially in smaller power systems and where, for example, wind conditions are good, because of the high cost for small-scale conventional generation. However, some of the challenges with high renewable penetration is that the systems can become rather complex and might include rather unusual equipment. Complexity and unusual equipment will not only affect the installation cost, but also impact maintenance.

Frequency-controlled dumpload is an example of one type of special equipment that can be found in some systems with high wind penetration. Such a device is typically designed to absorb excess generation from wind turbines and to control the grid frequency.

With small-scale renewable energy systems becoming mainstream and more mature, packaged solutions will reduce the complexity issue. What once was unusual equipment will become the new standard. Many more people will learn how to maintain them.

5.2.1.3. New developments: fuel cells in hybrid systems

One exciting development during recent years is the use of fuel cells and related technologies, which can in some situations bring special advantages, especially for hybrid power systems. One such situation is when a backup power source is desired, which is only seldom used. Some well-designed fuel cell systems apparently are able to stand still for years, without exercise runs or replacement of fuel, and still start reliably when needed. Another interesting situation is long-term energy storage when both heat and power are needed (the electric efficiency of e.g. an electrolyzer-hydrogen storage-fuel cell system is so low that it is not as attractive to use to replace a traditional battery if only electricity is of interest). The operation of fuel cells is usually quieter than other alternatives, which can have several advantages, including the ability to place the fuel cell where more of the heat can be utilized. Fuel cells are built for different fuels, such as hydrogen, methanol or biogas. In some places, this can enable 100% renewable energy supply at a lower cost and with much less environmental impact than previously possible. Two case studies with different fuel cell systems can be found in Chapter 6.

5.2.2. Hybrid systems for water supply

If both electricity and water supply are needed on the site, one solution is to create a hybrid system for electricity supply and add electric water pumping or desalination equipment as deferrable loads (loads that can be turned off for a short period).

Wind-diesel systems have also been designed which can use waste heat for desalination by distillation. The waste heat can come from diesel engines or surplus wind energy. This is an alternative mainly for sites with low to medium wind penetration, because it provides a good way to use the "free" waste heat from the engines. However, converting wind-generated electricity to heat for distillation is not an energy-efficient way to desalinate water.

If there is much surplus wind energy, it is more efficient to use the electricity to operate, for example, a reverse osmosis desalination plant.

For places that only need water pumping, hybrid systems are also available. One such system is described in the following example.

5.2.1.1. Example of a commercial product for water pumping

The Grundfos SQFlex system is a water-pumping package for remote sites that have no access to grid power. It can be powered by wind, solar PV or an engine-driven generator.

SQFlex Combi is the hybrid version, which allows both solar and wind input simultaneously (Fig. 5.7). Electric transmission of power is used, which means that the wind turbine can be placed in the best wind location, up to several hundred meters away from the pump.

There are no batteries in the system, just a few simple components, which include pump, breaker box, wind turbine and/or PV modules. Consequently, the system will pump water only when the wind is blowing or the sun is shining. The varying water flow can be evened out with a water reservoir, and a special level switch cuts out the pump when the reservoir is full.

Applications of the SQFlex system include, for example, livestock watering, crop irrigation and residential water pumping. The system uses submersible pumps and is suitable for wells with a depth of up to approximately 120 meters. Depending on local conditions, it can supply up to 150 m³ water per day.

One type of wind turbine that has been used in the system is Southwest Windpower's Whisper H80, a machine with 1-kW rated power, 3-meter diameter and standard tower heights up to 24 meters.

Optional components include the CU 200 combined control and status unit, which communicates with the pump via the power line. Grundfos also offers tools to help determine the system configuration needed.

Compared with conventional wind-electric pumping systems, the advantages of the SQFlex concept can be summarized as follows:

o   Very flexible power supply – different sources of power can be connected simultaneously and the motor can operate on 30–300 V DC or 90–240 V AC.
o   Problems regarding polarity have been eliminated – regardless of how you connect the SQFlex, the pump will always turn the right way.
o   "Plug and play" approach.

Figure 5.7 Wind-PV system for water pumping in Mozambique, 2002.

Source: Jan Lyngholm, Grundfos.

5.2.3. Hybrid system design and implementation

Designing and implementing a hybrid system can be a very demanding task, and is often carried out by a professional. Since a hybrid system involves different technologies, it is important to understand all the technologies that may be used on the site. However, the technical development and experience gained during recent years has made it easier to have success with hybrid power systems. When the book *Wind-Diesel Systems* by Hunter & Elliot was published in 1994, wind-diesel systems were mostly at the stage of research or demonstration, but today commercial hybrid system packages are available.

The amount of work that is used for project preparation naturally depends on the size of the project. Risø's report no. 1257, *Isolated Systems with Wind Power: An Implementation Guideline*, gives guidance for large projects (the report states that it is intended for sizes of 30 kW–10 MW).

For smaller systems, a simpler approach must normally be used. The first step in hybrid system design is often to make a site survey (also called site investigation) of the place in question. A checklist can be used to record the findings of the site survey. As an example, the checklist for design of a household-size system may include the following topics:

* Electric loads. This includes present and expected future load, opportunities to use energy more efficiently, possibility of deferrable loads (demand side management), seasonal influence, etc. A load profile (see the example in Figure 5.8) is often used as an input in simulation software, but a real measurement is not always available.
* Status of present generation, including electricity distribution system.
* Cost of fuel, including transportation.
* Conditions for renewable energy, such as wind, solar and hydro. Crude estimations of wind speed often have to be used, because a professional wind assessment can cost more than a small wind turbine. Noting local obstacles for wind and sunshine is important.
* Environmental protection. Are there any special local issues to consider?
* Siting considerations, for example access and soil type.
* Climatic and other conditions that can affect specification of equipment, such as ambient temperature, snow, risk of icing, windborne contaminants (e.g. sand), corrosivity of the environment or other special risks (e.g. rodents or termites).
* Human resources. Can local people carry out the maintenance needed?
* Logistics. Especially important on remote locations. How will equipment and spare parts be transported to the site?

Because there are many factors to consider when designing a hybrid power system, professionals often use computerized calculation programs to determine system configuration. These calculations use input data from, for example, the site survey and wind maps. The output is a preliminary system configuration and calculations of costs, fuel savings, etc.

HOMER is an example of such a program for simulation of power systems, which can be used not only for hybrids. Another example is SMA's Sunny Design. Two open-source alternatives are Opti-CE and micrOgridS [3].

When buying a hybrid system, it is often better to look for a turnkey package rather than individual components, to reduce the risk of ending up with equipment that doesn't function well together.

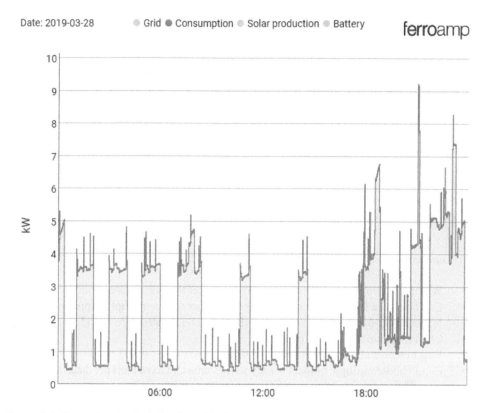

Date: 2019-03-28   ● Grid ● Consumption ● Solar production ● Battery   **ferroamp**

Figure 5.8 Example of a daily load profile from a Swedish home with heat pump, taken from the Ferroamp Portal.

Some related IEC specifications and standards are IEC/PAS 62111 (Specifications for the use of renewable energies in rural decentralized electrification), IEC 62257 (Recommendations for small renewable energy and hybrid systems for rural electrification), IEC TS 62898-1 "Guidelines for microgrid projects planning and specification" and IEC TS 62898-2 "Guidelines for microgrid operation".

5.2.4. Operational issues

Making a hybrid power system work in the long run is not just about buying and installing the equipment. Because a hybrid system involves several technologies, it is also more complex and the operational issues are even more important than for less-advanced solutions.

As with any other technical system, there must be adequate documentation of the installation. Otherwise troubleshooting can be difficult. Backup of, for example, software and parameter settings is essential.

Operators and maintenance personnel must have proper training. Unqualified personnel can sometimes cause more problems than they solve.

Funds must be allocated for maintenance and repair, for example, to replace a worn-out battery bank or hire a specialist for troubleshooting.

5.2.5. Open systems

Since hybrid systems are about combining different generating technologies, an important aspect, especially in larger systems, can be an "open systems" approach to avoid being tied to a single company, for example for replacement parts, maintenance and future expansion.

Openness has for some years been a general trend in control systems for many other purposes. For example, the LonMark organization was formed in 1994 and has a mission to enable the easy integration of multi-vendor systems based on LonWorks networks. Because control systems are part of most hybrid power systems, a similar trend could be expected in this area too. LonWorks is a very suitable technical platform for this, which also enables convenient integration with building automation, automatic meter reading, outdoor lighting control, demand side management, etc. Especially the possibility of power line communication (optionally with repeating) is of interest, for example, for load control, where other forms of communication might not be practical.

LonMark's nine questions for finding a long-term solution:

- Will my system be open to competitive bids after the initial installation?
- Can I install a system with multiple user interfaces from multiple suppliers?
- Is there built-in security at the low-level network-infrastructure level?
- Can I maintain my system by myself?
- Will I receive all the tools I need to fully maintain my system?
- Can I choose multiple bidders for my subsystems and have their products all integrated into one enterprise system?
- Is my system designed for only a small portion of my integration needs, or can it work with all of the components?
- Can I select products from multiple vendors and distributors and not be locked into a single vendor or source?
- Will all of the products that I select be guaranteed to work on the same network infrastructure?

In addition to the control aspects, the modularity, electric power interfaces, mechanical interfaces and contents/quality of maintenance documentation provided are essential to examine to avoid being locked in to a particular supplier.

5.2.6. Control and monitoring

Devices such as charge regulators and inverters usually have a built-in functionality for control, which can be sufficient in some systems. It has already been mentioned that in battery charging systems, the DC output of different charge controllers can often simply be connected in parallel. Settings need to be checked, of course, so they, for example, curtail output during high battery voltage at slightly different levels (so the one with a higher setting will finish the charging). Some solar charge regulators also have a separate DC output to load, with a low-voltage cut-off, to avoid discharging the battery too much. Advanced inverters, such as the Xtender series from Studer Innotec, can be configured to automatically start and stop a backup source (e.g. a diesel genset) or some load, based on battery voltage. Fuel cells for these applications usually have a start/stop control built in.

Some devices are also equipped for on-site and remote monitoring, or the manufacturer offers such options. Manufacturers often have web portals, where data are logged and displayed to the user (the load profile in Figure 5.8 is from such a portal). This can be sufficient, especially when all related equipment is from the same manufacturer.

For example, when additional functionality is needed or equipment from different manufacturers is used, a separate control and monitoring system may be desired.

TEROC has developed free open-source software for control and monitoring of hybrid systems not tied to a particular manufacturer. For example, software for "House Unplugged" (a case study in Chapter 6) is available, which includes a smart home system with basic load control. It is based on using a Wago PLC as main controller. TEROC also publishes software for a more modular system that allows distributed control, including automatic genset control (that also can be based on the health of the battery bank), communication drivers/templates (for equipment such as inverters and weather stations) and load control, which can be based on LonWorks or other communication. (See website links at the end of the chapter to download.)

Load control can have many functions. One is automatic load-shedding when there is a shortage of power or energy. Another is to automatically turn on loads when there is a surplus of renewable energy. This can enable some loads to be shifted in time, so renewable energy will be utilized better and there will be less dependence on electricity storage. Ramping up and down of some loads can also be an important function of load control, for example when large in-rush currents should be avoided.

5.3. Internet websites

RETScreen International, www.retscreen.net

HOMER, www.homerenergy.com

Opti-CE, optice.net

LonMark, www.lonmark.org

Templates and software for control and monitoring of hybrid systems, by TEROC, www.teroc.se/web/page.aspx?refid=77 www.teroc.se/web/page.aspx?refid=70

PowerSpout resources, for example for small hybrid systems, including hydropower, www.powerspout.com/pages/resources

5.4. References

[1] *Mini-grid Design Manual*, ESMAP Technical Paper 007, UNDP/World Bank, 2000. Available from: https://www.esmap.org/node/1009 [Accessed 2019-04-02]

[2] IEA-PVPS T11–07:2012.

[3] S. Berendes, Hybrid Mini-Grids Sizing with Microgrids, *Windtech International*, vol. 15, no. 5, July/August 2018.

Chapter 6

Case studies

Eight case studies explain different energy systems, which are at the forefront of using new technologies related to small-scale renewable energy.

6.1. Hybrid system with direct methanol fuel cells

Boliden Aitik is Sweden's largest open-pit copper mine, located near Gällivare. Large tailings ponds are used to store sand and water from the processes at the mine. The ponds are surrounded by dams. Although the risk of a dam failure is very small, there is such a risk. Boliden Aitik has a high-tech surveillance system, which uses sensors to monitor dam safety in real time. In connection with the negotiations for a new license for the operations at Aitik, the company was instructed to install a warning system for the public in the event of a dam break. This system has been in operation since February 2018. In case of an alarm, a sound signal and spoken message will warn the public in the surroundings, so they can reach a safe place.

The warning system consists of four siren stations with sound horns from Toleka mounted on Scanmast towers, placed in four strategically selected locations. One of these places is several kilometers from the nearest electricity connection point. For this site, electricity is provided by a combination of solar panels and direct methanol fuel cells (DMFCs). These fuel cells are made by the German company SFC Energy and supplied by their partner Awilco.

Solely using solar power is not an option for all-year power supply in this application, which is located north of the Arctic Circle. The hybrid system with DMFCs have been found to be a good solution, even though outdoor temperatures can be as low as about −35 °C. The conditions for small wind are not considered suitable on this site, which is surrounded by tall trees.

Some of the key components of the electricity supply are shown in Figure 6.1. Outside the shelter are four solar PV panels. Two panels are connected to each ProStar MPPT 40M-charge regulator from Morningstar, which charge the batteries as shown in Figure 6.2.

All loads use DC. However, the sound horns use 48 V and the communication equipment, including a 4G modem, uses 24 V. Connecting 24-V equipment to a 48-V system could be done using a DC/DC converter, but in this case the 24-V equipment is instead connected to two of the four 12-V batteries. To avoid imbalances, battery equalizers were first used, as shown in Figure 6.2. However, it has been found that the equalizers are not necessary in this case.

The typical load is roughly 20 W. During alarms and testing (done twice a day), the load will be higher.

Only when the battery voltage drops below a certain level will the fuel cells start to perform a charge cycle. At low temperatures, the fuel cells will also use some methanol for internal heating. The heat from the fuel cells helps to keep the temperature up inside the

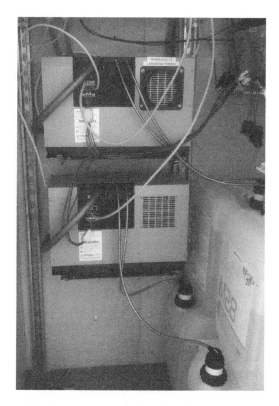

Figure 6.1 View of the inside of the shelter, with the two DMFCs (EFOY Fuel Cell Pro 2400 Duo set 12/24 V DC 110 W) in the center. The four methanol fuel cartridges of 28 liters each are seen on the right-hand side, below the batteries.

Source: Åke Karlberg.

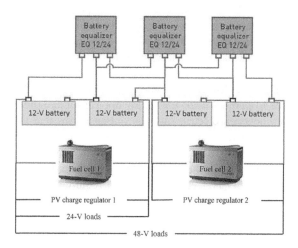

Figure 6.2 Simplified schematic of the electrical system in the shelter.

insulated Mava shelter during winter, which is good, especially for the batteries. The air exhaust from the fuel cells is vented mainly to the outside.

The warning system is monitored, including measurements such as fuel level, from remote using Boliden Aitik's ordinary control system from ABB (which is also used to trigger an alarm in case of a dam break). Fuel cartridges need to be changed about once a year. The main maintenance is that during some weather conditions, ice needs to be removed at the water outlet, approximately every two weeks. This could probably be improved with a different design of that detail.

The system is CE marked. Project management was done by the consulting firm Lividus and construction mainly by the company Eitech.

Boliden uses DMFCs for other applications as well, and are satisfied with their operation.

6.2. Self-sufficient home with pellet-driven CHP

The Austrian company ÖkoFEN is an established manufacturer of wood pellet heating systems. They have also developed the first CHP solution with a Stirling engine that is commercially available for ordinary homeowners. Their first product of this type was the Pellematic Smart_e, which provides 0.6–1 kW electric power output and 9–16 kW heat, and has an integrated accumulator tank holding 600 liters of hot water. The second generation of their CHP technology is the Pellematic Condens_e, with a similar output, but no integrated accumulator tank (Fig. 6.3). It is actually possible to start with their "e-Ready" Condens pellet boiler, for only producing heat, and later add the Stirling engine and generator to achieve a CHP solution – the Condens_e.

Thus, a stepwise approach to independence is possible. Under central European conditions, according to ÖkoFEN, it is typically possible to achieve about 30% electricity independence by a solar PV system only, 70% by adding energy storage in the form of a battery and 100% by adding the Stirling CHP.

In the village of Lembach im Mühlkreis, the founder of the company, Herbert Ortner, has equipped his own house with a complete package that enables the house to function off-grid. The main components are the Condens_e, a separate accumulator tank, a pellet storage system, a LiFePO4 battery with 9.6 kWh useful capacity, PV panels and power electronics from Fronius. The fully automatic system has been in operation since December 2017.

All of the main components are in the basement of the house, except for the PV panels (Fig. 6.4). That works fine because of the clean, safe and low-noise level operation of the Condens_e. This way of placing the system means that almost all the supplied energy can be used with minimal losses. However, some care may need to be taken when placing the CHP unit, so vibrations from the Stirling engine are not transmitted throughout the structure.

The rather low ratio between electric and heat output means that the solution is mainly suitable where there is a low demand for electricity and a high demand for heat.

This house can operate both with and without the public grid and is therefore equipped with a hot water tub in the garden, where excess heat can be dissipated (e.g. when no public grid is available and electricity is needed and no heat required for hot water and heating). Normally the house is connected to the public grid, which enables selling surplus electricity.

Because of the "emergency power function" of the Fronius Sumo Hybrid inverter, with an additional Fronius Smart Meter and switchover, this inverter will control voltage and frequency in the house if the grid is disconnected. The Stirling generator is connected directly to the AC-bus of the house.

Figure 6.3 Product manager Miriam Gahleitner operates the control panel of the Condens_e. On the left hand side is the Fronius Solar Battery 12.0 cabinet, with its Fronius Symo Hybrid 5.0–3-S power electronics on the wall. The exhaust is on the right.

Source: S. Ruin.

Figure 6.4 The self-sufficient home of Herbert Ortner.

Source: ÖkoFEN.

ÖkoFEN sells the complete package for independence, including the main components mentioned previously. In other off-grid homes using their CHP unit, power electronics from Victron have also been used.

Because the Smart_e and Condens_e use a condensing boiler with very high efficiency, this makes up for the extra fuel for driving the Stirling engine, so the use of pellets (approximately 2.5 kg/h for a thermal output of 12 kW) is comparable to a standard pellet boiler of similar thermal output. ÖkoFEN requires that the pellets confirm to the ENplus-A1 standard (see: www.enplus-pellets.at/qualitaet/). In Austria such pellets presently cost about € 230/ton if purchased in bulk. This corresponds to a fuel cost of € 0.0511/kWh, assuming the theoretical case that there are no losses.

There are also investment and maintenance costs. The planned maintenance is done once per year, when the Condens_e must be inspected by a person trained by ÖkoFEN. At that time, cleaning of the boiler is done. Also, the ashes must be removed by the user two to three times per year. Unplanned maintenance must also be done by personnel trained by ÖkoFEN.

The Microgen Stirling engine, which is part of Smart_e (no longer available) and Condens_e, is completely sealed and actually has no planned maintenance. This type of engine is also used by other manufacturers of micro-CHP, such as Vaillant and Viessmann, using gaseous fuel.

A larger CHP unit with another type of Stirling engine, for 4–5 kW electrical output and 50–60 kW thermal output, is also being developed by ÖkoFEN in the Pellematic e-max project.

6.3. House Unplugged

"Hus Utan Sladd." which can be translated to House Unplugged, is a concept house located outside Sigtuna in Sweden, and was built by the company Sisyfos (Fig. 6.5). The idea is to show that an ordinary Swedish home of about 120 m^2 can function independently. The house was from the start built to be off-grid, although the grid is available on the site. It also has a recirculation water system, where the outlet from the Alnarp Cleanwater natural wastewater treatment system and collected rainwater is stored and used to flush the toilet and water the garden. However, water for drinking, etc., is from a local well. Water consumption is minimized also by using a recirculating shower system from Orbital.

The house is very well insulated, like a passive house. Because all appliances are selected to minimize energy consumption, the house needs some additional heating during the cold months.

On the roof and two vertical fences are solar PV panels with a total rated power of 19 kWp. On the side of the garage is a small wind turbine, "Windflower," from the company Windforce. It has a 2-meter rotor diameter and can supply up to 1 kW, but the wind conditions are poor on the site, in particular close to the ground. Such a small wind turbine is therefore not really useful here. Experience has shown that the PV and 12-kWh LFP battery is sufficient to run the house off-grid more than half the year, including the heating based on storing surplus PV power in an accumulator/hot water heater.

For the few dark months when that is not enough, a small-scale CHP solution was intended, but did so far not work well. Therefore, it has actually been necessary to connect to the grid during some periods. During those periods, the house still has grid backup functionality (in case of a power outage on the grid), and will keep running without interruption during a number of hours, depending on parameters such as load.

Figure 6.5 Inauguration of the house on 2 May 2016. The solar PV panels on the other side of the house are together with a battery in the garage powering the house and charging two electric cars, off-grid. The small wind turbine did not contribute with much this day.

Source: S. Ruin.

The electrical system is based on a Ferroamp's DC-nanogrid with 760-volt DC nominal voltage. Unlike ordinary Ferroamp systems, the first plan here was to make an "all DC" house, to avoid all conversions of power between DC and AC. However, because of the difficulty of finding suppliers for DC appliances, a compromise had to be made. When needed, an inverter is started that converts 760 V DC to AC, so ordinary appliances can be used. DC is still used as much as possible and is sufficient to run everything needed, including the heating, when no one is in the house. The other DC voltages used in the house are 350 V, 24 V and 12 V, plus USB outlets (5 V).

In the beginning, AC was used to charge electric cars, because that was the only available solution at the time. Now the plan is to install Ferroamp's bi-directional DC charger. That will enable more efficient and faster EV charging, and most importantly, the house to be supplied from an EV when needed, so-called V2H (vehicle-to-home). In combination with the energy system of the house, an EV would keep the house running during the darkest part of the year, provided that the EV is regularly charged somewhere else.

If this concept house is placed in a location with decent solar conditions every day and/ or better wind conditions, it will function independently all year-round even without CHP or V2H, which has attracted visitors from other countries. The lightweight construction materials used could also make it interesting, for example, for floating homes.

Today the house is used as an ordinary home, where the project manager, Jens Johansson, lives. The project is still ongoing. Investigation is under way to determine if there is an approved solution for using the collected rainwater for potable water.

Figure 6.6 Mirror with "head up" display, showing data from the house.

Source: Jens Johansson.

One of the special design features in the house is that, next to the entrance, a mirror with a "head up" display is placed, as shown in Figure 6.6. The house also has battery-less and wireless EnOcean sensors, connected to the smart home system.

Software developed by TEROC in the project for a Wago PLC and Raspberry pi, used for controlling and monitoring the house, is available as free open-source software (see link at the end of the chapter).

6.4.   Home with seasonal storage of hydrogen

In the Gothenburg area, the electrical engineer Hans-Olof Nilsson has built a unique house, with seasonal storage of hydrogen (Fig. 6.7). It is his own home, of about 500 m², where he lives and carries out a live test of the energy system that he has integrated himself. The electrical system is powered by the sun, and the house has run off-grid since 2015.

Approximately 23 kWp of solar PV panels are placed mainly on the roof, but also on parts of the façade. In addition, there are solar thermal collectors on the roof. The PV power is fed via Viessmann/SMA Sunny Boy inverters to a three-phase, AC-bus based "island system," where the voltage and frequency is controlled by a cluster of SMA Sunny Island battery inverters connected to lead-acid batteries with a total nominal capacity of 144 kWh. The battery alone can power the house for up to five days. The maximum possible electric power output of the system is 48 kW short-term. The power and energy capacity of the house is enough to also charge the electric cars of the household and those of some of the visitors.

When the batteries are above about 85% SOC, the 5-kW PEM electrolyser HyProvide P1 from GreenHydrogen in Denmark is started to produce hydrogen and oxygen from deionized water. Water comes from the city system and a deionizer is started automatically when the electrolyzer is started. The hydrogen is stored at 300 bar, after being further compressed in a two-stage metal hybrid compressor without moving parts. The compressor uses cold water from the well and hot water from the solar thermal collectors. In total, 5.5 kWh of electricity and one liter of water are needed to make one normal cubic meter (Nm^3) of hydrogen. The new hydrogen storage, which is under construction when we were visiting the house, consists of 55-liter gas bottles, with a total volume of 12 m^3 if they were filled with water. The hydrogen storage is the most expensive part of the system. All oxygen produced is released to the atmosphere. Because the electrolyzer will run during the sunny part of the year, when there is normally an abundance of heat, the waste heat of the electrolyzer is not used and is dumped into the energy well for the heat pump. At 30% battery SOC, the electrolyzer is stopped.

Figure 6.7 The home of Hans-Olof Nilsson in Angered. However, other houses in the residential area and the wind turbine in the background are connected to the public grid.

Source: G. Sidén.

About 3000 Nm³ of hydrogen can be produced during a normal summer. Of that, around 2200 Nm³ is needed for the house. The surplus of hydrogen could be used for fueling a hydrogen car. The maximum storage capacity is 3000 Nm³.

During the dark part of the year, the PS-5 fuel cell from Swedish company PowerCell uses the stored hydrogen and oxygen from the atmosphere to produce 5 kW of electricity and about the same thermal power, which is used to heat the house (Fig. 6.8). Each Nm³ of hydrogen can be converted to about 1.5 kWh if electricity and 1.5 kWh of heat. The exhaust from the fuel cell is water, which is not used (but would be possible to store to use later in the electrolyzer, e.g. in case of a water shortage).

The hydronic heating system of the house also includes a ground source heat pump (where the ground during the summer is heated by the solar thermal power), accumulator tanks (3 m³) and hot water heater for tap water (800 L). Low temperature is used for floor heating. Surplus heat will be used to melt snow and ice in the driveway. The heating system components mainly come from Viessmann, because they could offer a very complete solution. There is also a two–heat exchanger ventilation system in the house.

Figure 6.8 Hans-Olof Nilsson in front of the fuel cell.

Source: S. Ruin.

Power to gas installation keeps a family home and their EVs running around the year

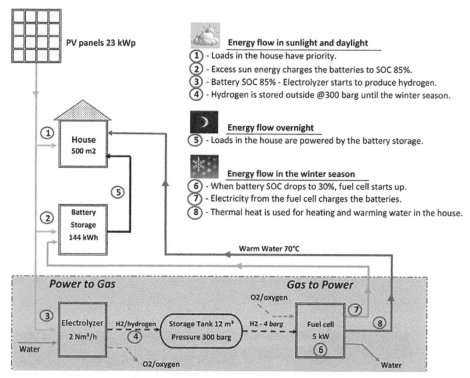

Figure 6.9 Simplified schematic of the storage system.

Most of the technical equipment is located in the large basement. Hydrogen detectors are used to detect leaks. Care is taken when placing the outlets of oxygen from the electrolyzer and possible hydrogen from the fuel cell, in particular to avoid the dangerous mixture of 4–75% hydrogen gas and air.

A simplified schematic of the storage system is shown in Figure 6.9. The entire system is automated and equipped with an advanced monitoring system. The collected data are an important base for the work Nilsson is doing in his company Nilsson Energy AB, to commercialize this type of system in other projects. He calls the solution RE8760.

For new systems, the electrolyzer can run with variable power and there might not be any solar thermal system. According to Nilsson, the system is suitable for scaling up, for example, to multi-family houses. He also works with solar-powered refueling stations for vehicles.

6.5. Farm with woodchip-driven CHP and solar PV

Andreas Huber runs a farm in the village of Sünzhausen, in southern Germany. Behind the electricity meter of the farm, he has installed a 120-kWp solar PV system and a CHP unit, rated 9 kW for electric power and 22 kW for thermal, which runs on woodchips. (On the farm there is also a 1.8-MW wind turbine, with another owner under a land-lease agreement and a separate grid connection at medium voltage 20 kV.)

The farm includes nine hectares of forest. Some of the wood harvested (usually during winter) there, which is less suitable for other purposes, is used to produce woodchips on site using a wood-chipper powered by a tractor. Then the woodchips are stored under roof and dried for usage the following winter. The CHP unit is an HKA 10 made by the German company Spanner Re2 (Fig. 6.10). Some of the heat produced by the CHP unit is used to dry the woodchips (to 15% water content or less) before they are fed to the wood gasifier, which is included in the HKA 10.

In the gasifier, the wood is converted to wood gas, which includes carbon monoxide. Because carbon monoxide is highly toxic, the plant is equipped with a warning system; it is placed in an outbuilding where no one lives. To have low moisture content in the fuel is important for the gasifier to function well – otherwise it needs to be cleaned from excessive tar.

After the wood gas has been cooled in a heat exchanger and filtered, it is fed to an internal combustion engine (a spark-ignition engine using the Otto cycle), which drives an asynchronous generator.

During the visit by one of the authors, the CHP unit actually produced 10 kW of electricity, which is a little more than rated. The output varies slightly depending on conditions. Heat from the CHP unit is transferred through a hydronic heating system to buildings on the farm.

Figure 6.10 A view of the inside of HKA 10. In the center is the engine (of Russian origin) with the generator connected on the right-hand side. The black cylinder on the top left behind the engine is the gasifier.

Source: S. Ruin.

The farm has an accumulator of 2500 liters and an additional heater of 100 kW, also using woodchips.

The process is automated, except for the preparation of the woodchips. In this case, the woodchips can be produced at very low cost on the farm, but also, should the woodchips need to be purchased, they are a cheap fuel.

Starting and stopping the CHP is normally done manually. It takes about 10 minutes for cold start and some seconds for warm start. Stopping the unit takes some seconds.

Planned maintenance includes lubrication, which needs to be done every 300 hours of operation. Every 900 hours and once every year additional maintenance is needed. So far, the owner has been able to do all the required maintenance himself – he is both an engineer and a farmer.

The farm pays more for buying electricity than it gets for selling excess production, so they try to use as much as possible of the electricity they produce themselves on the farm. One way they do this is by charging an electric car at the most suitable times. There is no automated load control system on the farm.

Through these investments in on-site generation of renewable energy, the farm is less vulnerable to increasing energy costs determined by others. Selling electricity also brings an additional stream of revenue to the farm.

The CHP unit normally runs continuously during the cold part of the year, from October to March. In addition, it runs during the nights during a few additional months. When there are long periods of good weather, typically in August and September, there is no need for heating, and the CHP is shut off.

However, all renewable energy systems on the farm are grid-following only, so they will stop automatically in case of a power outage on the grid. There is no emergency generator but the intention is to purchase one (or preferably an electric car which can have this functionality).

6.6. Municipal utility company using V2G and solar PV

Frederiksberg is one of the municipalities in Copenhagen, the capital of Denmark. At the headquarters of Frederiksberg Forsyning A/S, the municipal utility company, is the world's first commercial V2G (vehicle-to-grid) station next to the office building as well as a solar PV system.

The V2G station has 10 bi-directional chargers from Enel, which normally charge electric cars from the grid and can also use the batteries in the cars to supply power to the grid. Each charger is rated 10 kW, so a total of 100 kW is here at the disposal for V2G, which is in this case used for frequency regulation of the grid.

Batteries can be a fast-responding resource for frequency regulation, which is expected to require less maintenance than traditional solutions (such as fast control of the output of hydropower). Thus, the batteries in electric vehicles and the charging infrastructure for them can have a dual use – mobility and supporting the grid. However, for the battery to be able to provide power both to and from the grid, the state-of-charge (SOC) must be less than 100% during this time. Contrary to lead-acid batteries, which are usually considered to best be kept at 100% SOC, the modern lithium-based batteries often used in electric vehicles can actually fade less over time when being kept at a lower SOC.

Denmark has developed a trading system, where a service provider can make a certain power (kW) capacity for frequency regulation available on an hourly basis. Within this time frame, the V2G frequency regulation takes place by fast control (every second) of the charging/discharging of vehicle batteries. Thus, it is ideal for a fleet of service vehicles, which will much of the time be at this charging station and where the usage pattern can be well predicted (Fig. 6.11).

Here the cars are 10 Nissan e-NV200, which are used as service vans by the utility. Because of the small geographical size of this urban municipality, the range of this car is more than sufficient. Also Mitsubishi can deliver cars that can be used with V2G. Both these Japanese car manufacturers use the CHAdeMO quick-charging method.

Frederiksberg Forsyning has a contract with the company Nuvve, which in this case provides the V2G operation as a service, using Nuvve's own software. The software needs to know when the cars will be available for driving and which charging level is required at that time. Manual input of such parameters is used as a start. Later, the software will predict usage. There is also a buffer for unplanned car use.

By providing the V2G service for frequency regulation to the Danish transmission system operator (TSO), Nuvve is paid according to a dynamic pricing (which will vary over the day and year). The price is normally high during the evening and night, which is suitable considering the driving pattern of cars. In Denmark, Nuvve was paid approximately € 1400 per year and car during 2017, and expects to be paid more in the future. The profit is shared between Nuvve and their customer, in this case Frederiksberg Forsyning.

Today V2G is fairly expensive and mainly suitable for business customers, but in the future Nuvve plans to also provide "behind the meter services" that can be useful, for example, in homes.

Figure 6.11 For the users of the electric cars, the V2G system is as easy to use as an ordinary charging station.

Source: Frederiksberg Forsyning A/S.

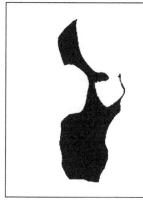

Samsø - Danish island in Kattegat
• Area: 112 km², 28 × 8 km, about 3700 inhabitants called *Samsingar* in 22 villages.
• The island, together with some smaller islands, is the municipality of Samsø, which belongs to the region of central Jutland.
• Ferry links: Hou, Jutland, to Saelvig, 21 km and Kalundborg, Zealand to Ballen, 35 km.
• Marinas: Ballen, Maarup and Langør.
• Important sectors for the economy:
- Agriculture. Basic products: Potatoes, beetroot, red cabbage and pumpkin.
- Tourism. There is a great interest for a renewable energy island. The marinas have many visiting boats.

Figure 6.12 The Danish island Samsø in 1997 won a competition to become a model society for renewable energy.

6.7.   Samsø – a renewable energy island

Samsø (Fig. 6.12) has a long tradition of being self-sufficient. When the island was to be electrified, it was done locally. At that time, direct current (DC) was used. The 110-volt direct current was not easy to transport long distances. The first electricity grid did not even cover the whole island. The power supply was made with five different power plants with DC generators powered by diesel engines [1]. The power plants were owned by local cooperatives. Eventually, electricity grids for alternating current (AC) began to be expanded in Denmark. Samsø was given the opportunity to connect to the AC–power grid in Aarhus County. But it took a long time before they did it. The cable to the mainland was expensive, and the inhabitants on Samsø had to pay for it themselves. Once the cable had been installed, residents had to pay for it through a surcharge on their electricity price. It took 10 years to pay off the cable.

The Ministry of Energy in Denmark presented an action plan, "Energy 21," in 1996 that stipulated the energy sector would go for a more efficient energy supply that would be less harmful to the environment and contribute to sustainable development for the Danish society. The report presented a target to reduce carbon dioxide emissions by 20% in 2020 compared with 1988. Renewable energy should be expanded by about 1% per year, which should yield an increase to 35% of the expected energy use in 2030. The government also wanted that in one local area, for example, an island, the energy supply should be converted to 100% renewable for all energy use, including the transport sector. Could Samsø become that area?

6.7.1.   Competition about 100% renewable energy

The Ministry of Energy announced a competition in 1997. Which local area or island could present the most realistic plan for a conversion to 100% self-sufficiency with 100% renewable energy? Four islands and a peninsula were invited to the competition: Læsø, Samsø, Aero, Møn and Thyholm. They had to develop a "master plan" detailing how they would carry out the conversion. In October 1997, Samsø was named the winner of the competition. The island's old tradition of self-sufficiency probably contributed to the victory.

The island would be a demonstration site for conversion to 100% renewable energy. It was about electricity, heat and transport that were based on, for example, wind power and solar energy. They should show the technical possibilities for switching to renewable energy and also make visible the economic and organizational opportunities that the transition provides for society. During a 10-year period, approximately 70 million Danish kroner would be allocated as financial assistance to the project, which would be completed in 2008.

One priority was to show how the consumption of energy could decrease in all sectors. Another priority was to show the degree of involvement from the local community. The business community, local authorities and organizations had to support the master plan to give it credibility.

The plan was to primarily use technical solutions that were based on available technology. For the financing and construction, the intentions were to use many different ways to organize, finance and own the new energy installations. From start they got a subsidy from the Ministry of Energy but most of the financing have come from local interests, the municipality, private inhabitants and organizations. The consultant and the Energy Academy have been important to plan and build the different plants.

6.7.2. The energy transformation

A consultancy firm, PlanEnergi, specializing in environmental and energy analysis, was engaged to make plans for the conversion. They inventoried how much arable land and forests could be used for energy production, which locations were suitable for district heating plants and where wind turbines could be located. Local energy use was also investigated and what climate emissions were generated. The share of renewable energy was 13% of energy consumption in 1997.

Plans for the future included opportunities to build 20 new wind turbines, a number of new district heating plants and many solar thermal and solar PV systems. Energy efficiency was an important part of the plans. For heating of buildings, use would be reduced by 30%, electricity consumption would decrease 25% and transport would use 20% less energy.

The city mayor was not so excited about the plans. Initially, the municipality would not participate in the funding. This was changed later. Through contributions from the energy authority, the energy consultant and a local enthusiast, Søren Hermansen, could be employed. The first half year was devoted to informing inhabitants about the vision at various meetings on the island.

Table 6.1 Renewable energy use and electricity trade at Samsø. The development during 16 years [3].

Year	1997	2003	2013
	GWh	GWh	GWh
Wind power	1.6	104.7	105
Solar heat/power		1.2	4.3
Heat pump			2
Bio fuel/energy crops			2
Straw	14.4	17.4	25
Wood/woodchips	1.5	4.5	24
Wood pellets/waste	0.4	6.1	7
Renewable energy	**17.9**	**133.9**	**169.3**
Electricity import/export	**27.2**	**−77.9**	**−77,1**

Figure 6.13 From the Samsø Energy Academy, work is being carried out on the transformation to a fossil-free island. It is also a center for information and education on the subject.

Source: G. Sidén.

An early project was a program for energy savings in the homes of older people. With a small contribution from the state, one could get a discount on insulation materials. Many home visits with information about efficient energy use were conducted and this gave a good dissemination of information about the project. Hermansen's local anchorage was important for the success. Locals felt confident about him and pride that Samsø had an important role.

The major work on the conversion has so far been the construction of wind turbines and the establishment of biofuel-fired heat plants. Table 6.1 shows that an import of electricity has been exchanged for a substantial export instead. The warming of the buildings on the island has also shifted from fossil fuels to bioenergy from straw and various forms of wood fuel.

In 2007, the target of 100% renewable energy was achieved [2], in a way. Still, a majority of the vehicles on the island still used fossil fuels, but that was compensated for by the fossil-free electricity that was exported.

6.7.3. *Samsø Energy Academy*

Samsø Energy Academy is a project organization that works with the consequences of climate change. In 2007, the building, which is a center for of the activities, was completed (Fig. 6.13).

It is a building with many features. Here the academy gathers and conveys knowledge from the island's many energy projects. The academy building is also a conference center for various activities where researchers and politicians discuss energy savings and new technologies. There is also an exhibition and energy school for the island's visitors and school students.

The Energy Academy has about 10 employees, who live on Samsø. The employees have professional skills in social development and sustainable solutions.

The intention was to create a place with attractive design for sponsors, users and locals to meet and work together. At the academy, meetings and gatherings are held for education, research, courses, seminars and exhibitions on energy and climate focusing on sustainability.

Samsø municipality also uses the academy and conducts activities with energy advice for companies and private tours, guided tours, workshops and seminars.

The building is located in a meadow with a view over the sea. The design and construction have their starting point in the place, in local building tradition and minimization of resource consumption. The field is an old seabed with high groundwater levels, so the house is built elevated on long concrete columns. The building is covered with zinc that is broken in the facades with black painted wood panels.

The house is designed as a low-energy building, with good insulation inspired by the German passive house concept. To minimize power consumption, the house has a daylight inlet and solar shading has been optimized. The house also has arrangements for natural ventilation. The fresh water consumption is low because rainwater is used for toilet rinse. Integrated in the roof's distinctive profile there is a hidden solar heating system and also solar cells.

Figure 6.14 Four 1-MW wind turbines in Samsø's landscape contribute significantly to the electricity supply on the island.

Source: G. Sidén.

6.7.4. Wind power

Wind conditions at Samsø are very good. As early as 1683, wind energy has been used on the island. A post mill that was built in the early 1600s was shipped from a neighboring island, Endelave. In 1899, a larger Dutch mill was built. Both still remain and are popular destinations for the many tourists.

It was natural that efforts to become a "Renewable Island" began with the construction of wind turbines (Fig. 6.14). In 2000, three wind turbines were built, each 1 MW, near Tanderup in Samsø. Two of the turbines are owned by farmers, the third by a local wind power cooperative. During the year, three more 1-MW units were built near Permelille. Here also, two are owned by farmers and the third by a cooperative. The third installation during the year was five 1-MW turbines near Nørreskifte. All are owned by farmers. A single 1-MW wind turbine produces annually what approximately 630 households consume annually. So on a yearly basis, these first 11 land-based turbines covered the entire electricity needs in Samsø, including needs for agriculture and other companies.

Ten off-shore wind turbines, 2.3 MW each, were built in 2002 south of the island. Five of the turbines were owned by Samsø municipality and the other five were owned by private investors. In November 2018 the municipality sold their turbines to a private company outside Samsø. They wanted to get capital for further investments in renewable energy projects.

The off-shore wind turbines deliver more energy each year than the total used on the island of fossil fuels for heating and transport, including ferries. The annual output is about 80 GWh.

In addition, there are a number of smaller wind turbines spread across the island.

The potential for further wind power is very high on Samsø. A wind investigation has shown that the potential for wind power is good both in the southern and northern parts of the island. There is also space for large parks for off-shore wind turbines in MW size in the shallow areas west and southwest of Samsø.

The municipality has the following guidelines for additional wind turbines:

- Wind turbines may normally only be installed at specified areas on the municipality's wind power map.
- Small household turbines with a maximum overall height of 25 meters can be installed outside the wind power areas in the vicinity of the property buildings, if they are based on a particular assessment that is compatible with the interests of the open landscape.
- Wind turbines must be placed and designed with the utmost account of the interests of the open landscape, and so that nearby disturbances are limited.
- The selection of wind turbine sites must follow the script "Guidelines and permits for wind turbines" as well as the instructions that are decided by the municipality.

Efforts to build wind power have been received very positively by island residents. Neighbors took part in planning, and could influence where and how many turbines were to be built. In total, € 40 million has been invested in wind power. A high percentage of the wind turbines are owned by locals. The turbines have been very profitable. The repayment period for turbines has been 6–8 years, which allowed the banks to grant the loans needed for the wind power cooperatives. The feed-in tariffs for wind power applied in Denmark have contributed to the profitability. For wind turbines connected to the grid in 2000, an electricity price of 0.43 DKK per kWh has been guaranteed for the first 22,000 full-time hours (approximately 10 years). In Denmark a law also exists that guarantees that 20% of a wind power project must be offered for local ownership.

6.7.5. Heat supply

In the 1960s, most of the houses on Samsø had oil-fired boilers installed. It was the cheapest and most comfortable way to heat up the houses. After the first oil crisis in the 1970s, oil prices began to rise more and more. Some farms installed a boiler that could be fired with straw. There was a surplus of straw, and not all of it could be plowed down in the fields. Some of it had to be burned to get rid of it. The straw was a local and cheap fuel that could be used more. In 1993, the first district heating plant on the island was built in Tranebjerg.

The plant uses straw as fuel (Fig. 6.15). Farmers who deliver straw to the plant get the ash back, so that they can return some minerals to their farmland.

Figure 6.15 Straw bales, which are harvested locally on the island, are the most important fuel for Samsø's district heating network.

Source: K. Sidén.

Today, straw delivers about 85% of the heat demand in Tranebjerg. A total of 263 homes, commercial companies, housing associations and institutions are connected to the heating plant, which has a power of 3 MW, and the annual supply is approximately 9500 MWh per year.

The district heating plant in Ballen/Brundby is also fired with straw; it is the only plant on Samsø that is 100% owned by the users. The plant supplies heat to 240 consumers in the two villages and is therefore located midway between the villages.

Even in Onsbjerg the district heating plant uses straw as fuel. It has 76 connected customers. Unlike the other heating plants, it is owned by a private company.

The district heating plant in Nordby/Mårup gets the heat from 2500-m² solar collectors and a 900-kW boiler that is fueled with woodchips. The woodchips come mainly from the forest in the south of the island. A total of 185 consumers receive heat from the plant.

Today, 75% of the homes on Samsø are heated with renewable energy. As they renovate district heating plants, the municipality wants a dialogue with the owners about more complementary solar heating.

6.7.6. Solar energy

Samsø is located in one of Denmark's sunniest areas, so it's natural that solar energy is an important part of the power supply (Table 6.2). Many of the municipal buildings have solar PV panels. In 2013, the municipality installed 4400 square meters of solar PV. They account for 80% of the municipality's own need for electricity.

There are many additional installations of solar PV on the island. Total production was 3.14 GWh, or 14.5% of the electricity demand, in 2015.

Figure 6.16 At the heating plant in Nordby/Maarup there are 2500 square meters of solar collectors and a cistern for storage of hot water. This system complements the heat from woodchips that are used as fuel.

Source: G. Sidén.

Table 6.2 Yearly solar energy supply at Samsø [4]

Supply	Capacity/quantity	Production
Solar PV	1.3 MW	3.1 GWh
Solar heat collectors, district heating	2500 m²	1.0 GWh
Solar heat collectors, individual	Approximately 517 m²	0.4 GWh
Total		**4.5 GWh**

For solar heating, the largest solar collector field on Samsø is located at the heating plant in Nordby/Maarup (Fig. 6.16). In addition to the solar collectors, a total of 2500 square meters, there is cistern storage for 800 cubic meters of hot water. Solar collectors provide a significant share of the district's heating needs. During the period May–August, the solar field covers most of the needs. At that time the demand for space heating is limited, but there are many summer houses and the residents there ask for much hot water for their showers. Often the boiler fired with woodchips is taken out of operation. Even in March, April and September, the sun contributes significantly to the supply. Annually the sun provides more than 21% of the district's heating needs. Denmark generally is the world leader in solar heating for district heating systems. About 25% of the needs are covered by solar heat. In several places there is large pit storage for solar-heated water.

6.7.7.   Efficient energy use

An important part of Samsø's energy project has been the efficient use of energy.

There are 2200 villas on Samsø, of which approximately 2000 were built before 1980. Of these, 1800 are considered potentially able to be made more energy efficient. About 300 households are heated with oil and 650 are connected to district heating; another 650 have other types of heating, including heat pumps or pellets/firewood. The total consumption of energy for heating on Samsø is 52 MWh per year. Bioenergy accounted for 35 MWh per year. Part of the challenge is to reduce heat consumption even in commercial buildings.

The plan for Samsø is that, by 2021, heating consumption in private housing will be reduced by 30% and in other premises by 5%. By 2050 the goal is to achieve a reduction in energy for heating in private housing by 35% and in commercial premises by 10%.

In the case of electricity, the challenge is that consumption of primarily heating will not increase during the period up to 2030 compared with 2010. It is estimated that electricity consumption at the 2010 level can be maintained through savings and advice, despite the expected increase in heat pumps.

For public institutions and companies, the goal is to save energy up to 30% by 2030 compared with consumption in 2010.

Samsø municipality, the Energy Academy and the pump manufacturer Grundfos cooperate in a demonstration site where it is shown that the consumption of electric energy for pumps can be reduced by 30% by changing old pumps for new, energy-efficient units.

Examples of planned actions are a campaign to show where consumption takes place which provides a good basis for action. Training of craftsmen will also be arranged, as well as work to replace compressors, pumps and electric motors.

The requirement for new buildings to have very low energy consumption (maximum of 15 kWh per square meter a year) also contributes to reduced electricity consumption.

6.7.8. Smart grid in Ballen marina

The consumption of electricity varies considerably at the marina in Ballen. When the tourist boats arrive in the summer, use increases sharply (Fig. 6.17). In autumn, winter and early spring, demand is significantly lower. Wind power on Samsø yields a big surplus over the whole year; even so, in the summer deficits can occur. An undertaking by the SMILE (Smart Island Energy System) project is planning to test a smart grid system.

In Ballen marina a number of solar PV panels will be installed in addition to the existing ones. They will produce electricity which will be stored in a battery bank. The stored electricity can then be used to meet the increasing needs of the boats in the evenings.

To be able to cope with a renewable energy supply, a form of leveling in the system has been proposed. Boat owners should be offered a benefit if they can be more flexible in their use of energy.

Today they pay a flat fee for access to the electricity grid, and then they can use as much as they want up to what the fuses can withstand. Part of the new proposal stipulates that they will get cheaper power if they wait to charge their batteries until the load on the electricity grid drops.

SMILE is an EU project that is not limited to Samsø. It also includes islands in Portugal (Madeira) and Great Britain (Orkney Islands). The development of smart grids is an essential precondition for the transition to an efficient and reliable energy system. The SMILE project will showcase nine different smart technologies on the three islands.

6.7.9. A fossil-free island

The next step in Samsø's development as an energy island is the plan to completely phase out all fossil energy use by the year 2030.

Ferry traffic is the largest user of fossil fuels. Biogas is a pure form of energy that can be extracted from the biological waste that is produced. It fits well for Samsø, which has a large agricultural sector. The production of biogas will play an important role if the ambitious goal is to be fulfilled.

In December 2017, the city council decided that a biogas plant could be built on Trolleborgvej at Samsø Syltefabrik's wastewater pool. This location is on a major road, which is close to a large part of the island's farming.

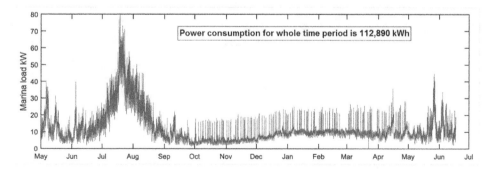

Figure 6.17 Consumption of electrical power in Ballen marina 1 May 2016–19 June 2017; use increases in July and August when tourist boats arrive [6].

The plant is still on the drawing board, but a finished facility will be able to receive manure slurry, energy crops and waste from Samsø Syltefabrik. They will also be able to use specially cultivated nitrogen-fixing crops and grass from the golf course – two of the projects within Samsø's pursuit of circular economy.

The facility will contain:

- Reception station with tanks and storage
- Agricultural biomass silos
- Digester and storage for after treatment
- Facility for gas cleaning, storage and upgrading

The cost for the biogas plant is estimated at 75 million DKK. A business plan for the plant is under preparation. Samsø Shipping Company will be a major customer and receive biogas for millions of euros each year.

In the years 2016–2017 an environmental impact assessment (EIA) was carried out, and environmental approval has been obtained. There will be a dialogue with potential investors and the largest purchaser of biogas, Samsø Shipping Company, by the end of 2018. The hope is that construction of the biogas plant will start in 2019–2020.

For cars, the plan is that 50% will be powered by electricity and renewable fuels by 2020. The municipality itself has switched all their cars to electric vehicles (Fig. 6.18).

Figure 6.18 Ninety percent of the cars owned by Samsø municipality are powered by electricity. The cars of the municipal home care are charged under a roof of solar PV panels in Tranebjerg.

Source: Samsø municipality.

To get the transports to fully switch to renewable fuels is the biggest challenge. Although the municipality has set up a number of charging stations around the island, the proportion of electric vehicles is not so high, even if it is among the highest in the country. Denmark has not had particularly large incentives for electric cars and the charging infrastructure is limited as soon as it comes to the mainland. However, electric cars are suitable for the driving pattern on a smaller island, which periodically has a large surplus of electricity from wind power.

For heavier trucks, buses and agricultural machinery, other fuels are needed. One possibility is that the biogas plant can cover a part of the needs. The goal is clear: All fuels for transportation on Samsø will be based on renewable energy.

Examples of actions taken into consideration are as follows [5]:

- More charging points for electric cars and the introduction of intelligent electricity meters
- Acquisition of electric buses and electric cars for taxi services
- Training of local mechanics for service of electric cars
- Establishment of biogas plants producing compressed biogas or liquid biogas for ferries, lorries, tractors and possibly, buses
- Establishment of filling stations for gas vehicles

The effort on Samsø to become an energy island has been done to create a long-term sustainable society and reduce the climate impact of fossil energy. But the investment has also had other effects. It has led to more jobs. The inhabitants on Samsø do not manufacture wind turbines but they can build and service them. The straw fuel will be grown, transported and harvested and the boilers for the biofuel will be taken care of and serviced. Samsø Energy Academy has also provided about 10 jobs.

The many visitors also provide employment for the tourist industry on the island. Often 30–50 people a day come to Samsø to get green inspiration. This has made the tourism industry an asset for conversion, especially as the energy center emphasizes the importance of having at least one overnight stay to fully understand the changeover. Those responsible for the island's tourism business have now changed their attitude. In the beginning, they said that the wind turbines would cause the end of tourism on the island; now the wind turbines are attractive elements in the tourist brochures.

6.7.10. Going further to energy independence

So far, 100% of renewable energy has been achieved by compensating for the fossil energy used with renewable electricity from wind power plants mainly. It works well as long as a large part of the energy supply in the surrounding area is fossil. The energy from the wind turbines and other renewable sources reduce the need for fossil energy. In the future, when fossil use decreases in general, a further developed strategy is needed. At Samsø, it is considered how to use locally even more of their self-produced energy. This means, for example, that they need to store electricity that can be used when the wind is not blowing. Strategies for this can be storage in batteries or production and storage as hydrogen.

Because the availability of electricity from wind power is so high, more electricity can also be used in heat pumps for district heating. Bioenergy from straw that is released can be used for biogas in the transport sector.

Smart-grid technology under development can also contribute. Energy use can be shifted to times when there is a surplus in the supply. For electric car owners, the car's battery can also be used for the general power supply.

6.7.11. Samsø has cheaper energy

The energy conversion at Samsø has always been characterized by the desire to use economically advantageous methods for the conversion. The new energy supply should preferably be cheaper for the residents. The people at Samsø do not pay less for electricity than other Danes, but many have invested money in the cooperative wind turbines, and they receive a refund that reduces the overall cost of electricity.

The use of straw and woodchips has also resulted in lower energy costs. The price of the oil replaced has increased on average during the conversion period. Even though the new facilities come at a price, the cost of heating is estimated to have decreased by 20% on average.

The fact that energy has become cheaper also contributes to the fact that energy conversion has received so much support from the population.

6.8.   El Hierro: self-sufficient island with pumped hydropower

The island of El Hierro (Fig. 6.19), belonging to the Canary Islands group, has been declared a "biosphere reserve and geopark" by UNESCO. The island has a unique natural and geological heritage characterized by its landscape and volcanic origins.

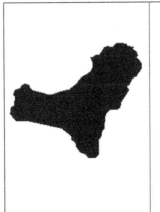

El Hierro, Canary Islands, Spain
- Area: 269 km², second smallest of the Canary Islands
- Coastline: 876 km
- Population: Around 10,700 inhabitants
- Capital: Valverde, three municipalities: Frontera, Valverde and El Pinar
- Highest point: Pico de Mapaso, 1501 m
- Like all of the Canary Islands, El Hierro is a popular tourist destination. The island receives about 20,000 visitors a year.
- There is a small airport - El Hierro Airport - at Valverde and a ferry terminal at Puerto de la Estaca, both with connections to Tenerife.

Figure 6.19 El Hierro, one of the Canary Islands, hopes to become self-sufficient with regard to electricity.

Since 1996, the island has had a sustainable development plan that ensures improvement of the population's level and quality of life and conservation of natural areas. The island is now striving for a 100% renewable power supply. It fits well with the UNESCO declaration, meaning that the environment has improved and carbon dioxide emissions are reduced.

6.8.1. Self-sufficiency with renewable energy

On isolated islands far from the mainland, it has been difficult to arrange a good and reliable electricity supply. Many islands have been unable to connect to mainland networks with conventional power plants. Instead, power commonly has been supplied by local generators running on diesel fuel, which must be transported to the island.

Oil prices have risen higher and higher since the oil crisis in 1973, and it has become an increasingly expensive method of power. Many islands have good access to renewable energy resources such as favorable wind conditions, waterfalls, good solar radiation and energy-rich waves. Of these, only hydropower can be controlled if it has dams or other storage facilities for the water. Electricity supply must have storage or production resources that can be controlled. Electrical energy must be generated exactly when someone needs it.

On the volcanic island of El Hierro, a megawatt-size electrical system has been developed that utilizes the good winds and controllability of hydroelectric power. Pumps powered by wind turbines and two artificial reservoirs and a hydroelectric power station can provide controllable power.

The project, "The Wind-Hydro-Pumped Station of El Hierro," whose cost amounts to € 64.7 million, is promoted by Gorona del Viento El Hierro, SA, with participation by the Island Council (66%), Endesa (23%), the Canary Islands Technological Institute (8 %) and the Autonomous Community of the Canary Islands (3%) [7].

The different parts of the system are shown in Figure 6.20. The facilities have been in operation since 2014, and the need for imported oil has been greatly reduced.

6.8.2. Desalination plant

Access to fresh water has a great economic and environmental importance for society, and the search for the valuable liquid has become an important process in many countries. The desalination of seawater has provided the opportunity for a new strategy for water supply, with the possibility of new abundant supply sources. The Canary Islands have a large population, a significant tourist industry and a great need for irrigation in agriculture. Therefore, desalination of water has been an important opportunity for water supply for the past 40 years. Between 3–25 kWh/m^3 are consumed in desalination by various methods. About 20% of the energy production in the Canary Islands goes to desalination and water distribution [8]. In the El Hierro hydropower system, the water used should be fresh water. A desalination plant is located adjacent to the lower basin. For desalination the electricity from wind turbines can be an important contribution.

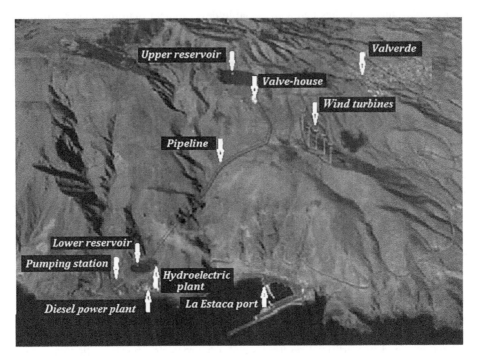

Figure 6.20 The various parts of the wind power and pumped hydropower plant at El Hierro.
Source: Courtesy of Gorona del Viento.

6.8.3. Wind power at El Hierro

The wind farm consists of five Enercon E-70 turbines (Fig. 6.21).

The total installed power is 11.5 MW. Electricity obtained directly from wind is used for the needs of the inhabitants, for the water desalination plant and for water pumping to the upper water storage.

One advantage of the Enercon E-70 solution is that wind energy converters are equipped with a special storm control feature. This slows down the turbine at 28 m/s and stops it completely at 34 m/s (10-minutes average value).

Thanks to this feature, the turbines can still operate at high wind speeds and do not stop abruptly. The reason that wind turbines are usually stopped at high wind speeds is that the total loads on the entire construction are lower for a motionless turbine compared with one in operation. A speed higher than 34 m/s very seldomly occurs so the energy losses are negligible.

These wind turbines with synchronous generator and full-converter electricity treatment also start to produce at a very low wind speed, 3 m/s. Thanks to that, the wind turbines operate a large share of the time. The turbine also has a very good efficiency, with an efficiency coefficient more than 0.50 at the most common wind speeds.

Table 6.3 contains the main data of the Enercon E-70 wind turbines, used at Gorona del Viento on El Hierro.

There is good access to high wind speeds at El Hierro. The average wind over the year has been estimated to about 7.5 meters per second.

Figure 6.21 The five wind turbines in the hilly landscape at El Hierro have access to high wind speeds.

Source: Courtesy of Gorona del Viento.

Table 6.3 Main data of Enercon E-70 E4 wind turbines [9]

Parameter	Unit	Value
Rated power	kW	2300
Hub height	meter	64
Rotor diameter	meter	71
Number of blades		3
Rotational speed (variable)	rpm	6–21
Cut-out wind speed	m/s	28/34

6.8.4. Hydropower with storage

The pumped hydropower project consists of the following parts:

- The upper reservoir is located on the La Caldera crater at an altitude of 715 meters above sea level (Fig. 6.22). Originally, it had been planned for 550,000 m^3, but since weaknesses of the bedrock were observed, the maximum capacity was limited to 380,000 m^3. It is sealed with 2-mm thick PVC membranes, which can be repaired under water.
- The lower reservoir is 56 meters above sea level near the Llanos Blancos diesel power plant. The reservoir has a capacity of 150,000 m^3 and is also sealed with PVC membrane.
- The water flow to the hydro turbines takes place through a 2350-meter tube, 1 meter in diameter.

Figure 6.22 The upper water reservoir on the La Caldera crater at El Hierro.
Source: Courtesy of Gorona del Viento.

- The pumping station, which is installed in a new building, consists of two pumps, each 1500 kW, and six pumps, each 500 kW, for total power of 6 MW.
- The hydropower plant consists of four Pelton turbines of 2.83 kW, which give total power of 11.32 MW. The maximum flow for electricity generation is 2.0 m³/s and the drop height, the head, is about 655 meters.

Based on potential energy, the power for a hydropower plant can be calculated using the following formula:

$$P = \eta \cdot Q \cdot H \cdot g$$

where: P = power, η = efficiency, Q = flow rate, H = head, g = acceleration due to gravity
When Q is in cubic meters per second, H in meters and g = 9.81 m/s², then we get P in kilowatts.

With an efficiency of 0.80, a head of 655 meters and the flow of 2.0 m³/s we get the maximum power for hydropower.

$$P = 0.8 \cdot 2 \cdot 655 \cdot 9.81 = 10{,}281 \text{ kW or approx. } 10.3 \text{ MW}$$

The demand is between 4 and 7 MW so all the island's power needs can be fulfilled. With a full water flow, 2 m³/s, the hydro plant can run for about 21 hours to fill the lower reservoir, but the needs are lower most of the time. The electricity generation at a full filling of the lower reservoir will be about 216 MWh. The total consumption in 2015 of 44.6 GWh gives an average daily use of 122 MWh, so in periods with very low winds the hydropower plant can produce electricity for about two days. The storage is not a seasonal storage, but rather storage for some days. If they can use a full upper storage of 380,000 m³, the time will be longer, but then a lot of fresh water must flow into the sea. Because there is much demand for fresh water and energy for desalination, it is not realistic to do that currently.

6.8.5. Results achieved so far

The aim of the project to become self-sufficient with 100% renewable energy has so far not been achieved, but the percentage has increased with every operating year. One reason for this result is that it is difficult to achieve 100% when wind power production varies so much during the year (Fig. 6.23). The monthly generation in June to August is more than double that of October to December. The energy storage system is not large enough to store the needs for late autumn.

A first milestone for the system was achieved in August 2015 when the entire island's need for electricity was met without diesel power for four hours. Following that, the periods of 100% renewable electricity supply grew longer. On 25 January 2018 and for 18 consecutive days, El Hierro's wind and hydropower plant supplied the entire island's demand for electricity. In July 2018, a total of 96% of the electricity needs were met by wind and hydropower.

For all of 2018, 56% renewable energy was achieved. Figure 6.24 illustrates the steady rise that occurred every year for the portion of wind and hydropower delivered to the electricity grid at El Hierro.

With its energy solution, El Hierro has become a good example of self-sufficiency with renewable energy; this solution can be used in many isolated areas around the world.

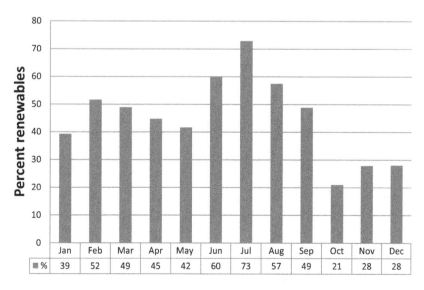

	Jan	Feb	Mar	Apr	May	Jun	Jul	Aug	Sep	Oct	Nov	Dec
%	39	52	49	45	42	60	73	57	49	21	28	28

Figure 6.23 Generation from wind and hydropower; average values for the second half of 2014 to 2018 [10].

With the current facility, the goal of 100% renewable energy can be difficult to achieve, as already mentioned. A significantly increased storage capability is needed to cope with the low winds of late autumn. Alternatively, other renewable energy sources that are independent of weather conditions can be installed.

An alternative could be a biogas plant with good storage possibilities for the biogas. In many countries, for example Germany and Denmark, electricity is generated in biogas power plants.

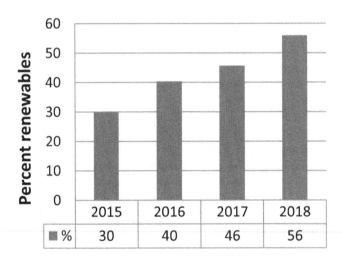

Figure 6.24 Percentage of renewable energy in the electrical power supply at El Hierro. In 2015, only figures for the months of July to December were available [10].

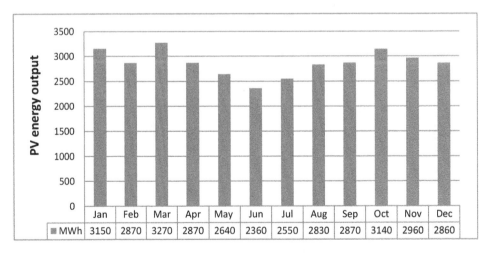

Figure 6.25 Calculated generation from a 20-MW photovoltaic power station at El Hierro mounted with a slope angle of 55 degrees.

Source: Calculation with PVGIS 5 [11].

Another option is to utilize the geothermal energy that can be assumed to exist on the volcanic island. However, there are no reports that geothermal resources for electricity production have been identified at El Hierro. That may be because no such investigations have been made. However, the prospects that there are some geothermal resources must be good. In Iceland, 29% of the electricity is taken from geothermal power plants (2013), even though the country has very large access to hydropower.

Another possible supplement is, of course, solar power. Figure 6.25 shows what the production would be monthly for a 20-MW solar PV plant located near Valverde. It could consist of 5128 pieces of 390-watt solar modules, each with a surface area of 2.1 square meters. A suitable location is a south-facing hillside. The total area of the solar panels would be 10,800 square meters. They should have a slope of 55 degrees. The large slope increases production at the beginning and end of the year (October–December) when the need for more renewable energy is largest. During the warmest period, production will be lower, but then the needs will be better covered by wind power.

Table 6.4 shows the monthly total production of electricity in 2018 and renewable production from wind and hydropower. If the calculated production from the solar plant is added, there would have been a surplus every month. On average, the surplus was 34%. Of course, a part of this can be lost through water pumping, but the final result would still be close to 100% renewable electricity.

The energy storage in reservoirs and hydroelectric power is very important for this solution with solar power. It enables energy to be stored from the daytime, when the solar radiation is intense, until evening and night. There are also other possible options for short-term storage. One prerequisite, of course, is that one can create a stable electricity grid independent of the diesel power plants.

Electricity from photovoltaics has been considered to be expensive. However, prices have changed much the latest years. According to *PV Magazine International* [12], the price for PV modules, mainstream-type panel, was as low as € 0.27 per watt in December 2018. For the whole 20-MW station, the cost will be € 5.4 million for the solar panels. The cost for a complete station can be around € 15–18 million.

Table 6.4 Generated electricity at El Hierro for the year 2018 and the impact from a 20-MW solar PV station, estimated by calculation with PVGIS [11]

	Total generation MWh	Renewable generation MWh	Calculated new solar PV MWh	Total renewable MWh	Surplus %
Jan	3601	2365	3150	5515	53
Feb	3228	1836	2870	4706	46
Mar	3461	1700	3270	4970	44
Apr	3365	2362	2870	5232	55
May	3625	2124	2640	4764	31
Jun	3837	2469	2360	4829	26
Jul	4089	3899	2550	6449	58
Aug	4275	3021	2830	5851	37
Sep	4063	2303	2870	5173	27
Oct	3912	966	3140	4106	5
Nov	3324	997	2960	3957	19
Dec	3447	999	2860	3859	12

6.9. Internet websites

Hus Utan Sladd (in Swedish), husutansladd.se

Download the free open-source software House Unplugged, by TEROC, www.teroc.se/web/page.aspx?refid=77

Samsø Energy Academy, energiakademiet.dk/en/

Energies, El Hierro Renewable Energy Hybrid System, G Frydrychowicz-Jastrzębska-2018, https://www.mdpi.com/1996-1073/11/10/2812/htm

ResearchGate, Sustainable energy system of El Hierro Island, R. Godina et al, Conference Paper, 2015, tinyurl.com/y5nub2tl (https://www.researchgate.net/publication/282575056_Sustainable_energy_system_of_El_Hierro_Island)

6.10. References

[1] Interview with Søren Hermansen, CEO, Samsø Energy Academy, 2018-11-16.

[2] P.J. Jørgensen, S. Hermansen, *Samsø – A Renewable Energy Island, 10 Years of Development and Evaluation*, Plan Energy and Samsø Energy Academy, 2007.

[3] B. V. Mathiesen, et al., *Samsø Energy Vision 2030*, Aalborg University, 2015.

[4] EU-Project, *Smile, Smart Island Energy System*, Deliverable D8.1, 2018.

[5] Samsø Kommune, *Kommuneplan*, 2017.

[6] EU-Project, *Smile, Smart Island Energy System*, Deliverable D3.1, 2017.

[7] Endesa, "El Hierro, an example of sustainability". [Online.] Available from: https://www.endesa.com/en/projects/a201611-el-hierro-renewable-sustainability.html [Accessed: 2019-05-30].

[8] S. Suárez, *Strategies for Promoting RES in Islands*, Canary Islands Institute of Technology, 2013. Available from: https://www.slideshare.net/UNDPhr/el-hierro-100-res-island-en, slide 12

[9] Enercon Product Overview, 2015. [Online]. Available from: https://www.enercon.de/fileadmin/ . . . / Medien. . . /ENERCON_Produkt_en_06_2015.pdf

[10] R. Andrews, "El Hierro Fourth Quarter 2018 Performance Update", *Energy Matters*. [Online]. Available from: http://euanmearns.com/el-hierro-portal/ [Accessed: 2019-03-30].

[11] PVGIS – Photovoltaic Geographical Information System, JRC, European Commission, 2019. [Online]. Available from: http://re.jrc.ec.europa.eu/pvgis.html [Accessed 2019-05-28].

[12] PV Magazine International, "Module Price Index", 2018. [Online]. Available from: https://www.pv-magazine.com/features/ . . . /module-price-index/

Index

Note: Numbers in italics indicate a figure and numbers in bold indicate a table on the corresponding page.

Printed and bound by CPI Group (UK) Ltd, Croydon, CR0 4YY

23/10/2024

01778253-0002